FRONTIERS IN ELECTRONICS

Selected Papers from the Workshop on
Frontiers in Electronics 2011 (WOFE-11)

SELECTED TOPICS IN ELECTRONICS AND SYSTEMS

Editor-in-Chief: **M. S. Shur**

---

*The complete list of the published volumes in the series can be found at
http://www.worldscientific.com/series/stes

FRONTIERS IN ELECTRONICS

Selected Papers from the Workshop on Frontiers in Electronics 2011 (WOFE-11)

San Juan, Puerto-Rico   18 – 21 December 2011

Editors

Sorin Cristoloveanu
IMEP, INP Grenoble – MINATEC, France

Michael S. Shur
Rensselaer Polytechnic Institute, USA

 **World Scientific**

NEW JERSEY · LONDON · SINGAPORE · BEIJING · SHANGHAI · HONG KONG · TAIPEI · CHENNAI

Published by

World Scientific Publishing Co. Pte. Ltd.

5 Toh Tuck Link, Singapore 596224

USA office: 27 Warren Street, Suite 401-402, Hackensack, NJ 07601

UK office: 57 Shelton Street, Covent Garden, London WC2H 9HE

British Library Cataloguing-in-Publication Data
A catalogue record for this book is available from the British Library.

Selected Topics in Electronics and Systems — Vol. 53
FRONTIERS IN ELECTRONICS
Selected Papers from the Workshop on Frontiers in Electronics 2011 (WOFE-11)

Copyright © 2013 by World Scientific Publishing Co. Pte. Ltd.

ISBN   978-981-4536-84-4

Printed in Singapore

PREFACE

The seventh in the series of the Workshops on Frontiers in Electronics – WOFE–11 – took place in San Juan, Puerto-Rico, in December of 2011 (see Figure 1). This meeting has attracted high scientific quality, visionary research presentations. As has become traditional for the WOFE conferences, research results were linked to applications ranging from defense and consumer electronics and optoelectronics to biotechnology and homeland security. Over fifty leading experts from academia, industry, and government laboratories presented their findings on the most recent and exciting developments in their fields, emphasizing future trends and emerging ideas. The scientists and engineers from all over the globe exchanged frank and sometimes rather controversial views on more or less expected directions of the electronics and photonics industry.

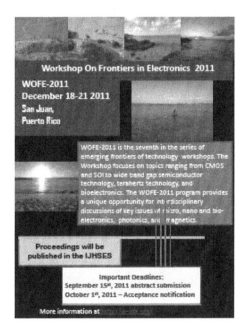

Fig. 1.

Traditionally, the WOFE Award Committee selects papers and posters for the Best Paper Award and Best Poster Awards. In 2011, the Best Paper Award went to E. Calleja, A. Bengoechea-Encabo, S. Albert, M.A. Sanchez-García, F. Barbagini, E. Luna, A. Trampert, U. Jahn, P. Lefebvre for their presentation on "Efficient phosphor-free, white light emission by using ordered arrays of GaN/InGaN nanocolumnar LEDs grown by Selective Area MBE." Byoung Hun Lee, Hyeon Jun Hwang, Eun Jeong Paek, Young Gon Lee, Chang Goo Kang, Sang Kyung Lee and Chun Hum Cho received the Best Poster Award for their work on "Applications of Metal/PVDF-trFE/Graphene Devices for Future Electronics." The second and third place in the Best Poster Competition went to Dongsheng Ma and Bingqing Wei for their paper "Carbon Nanotube Supercapacitor Based Reconfigurable On-Chip Energy Storage and Management for Autonomous Wireless Sensor Nodes" and to R. Palai and J. Wu for their paper entitled "Ferromagnetic Behavior in Ytterbium-doped and Ion Implanted GaN Semiconductor," respectively.

This issue includes the best papers of WOFE-11 invited by the Editors and down selected after the peer review process. This book is conceived to make available in the international arena extended versions of selected, high impact talks. The papers are divided into four sections: advanced terahertz and photonics devices; silicon and germanium on insulator and advanced CMOS and MOSHFETs; nanomaterials and nanodevices; wide band gap technology for high power and UV photonics. These key issues are in forefront of the micro-nano-electronics research and many papers in this book go well beyond the discussions of the original work giving a good overview of the field.

This book will be useful for scientists, engineers, research leaders, and even investors interested in microelectronics, nanoelectronics, and optoelectronics. It is also recommended to graduate students working in these fields.

This book is dedicated to the memory of Professor Dieter Schrode, our dear friend, who left us suddenly after completing his paper and before the publication of the book.

The next WOFE will take place in December 2013, also in San Juan, Puerto Rico. We welcome suggestions for topics to be addressed, special sessions, tutorials, and panels.

On behalf of the WOFE Organizing, Program, and Steering Committees, we would like to thank all the participants and especially the invited contributors to this issue for making WOFE–11 a successful conference. Our special thanks go to the Members of Organizing, Program, and Steering Committees, and to Session Organizers for their tireless work and inspiration. We also gratefully acknowledge generous support of this workshop by the National Science Foundation (award # 1138284, Program Director Dr. John Zavada) and by the Office of Naval Research (GRANT10951364, Program Manager Dr. Chagaan Baatar.)

EDITORS
Michael Shur and Sorin Cristoloveanu

CONTENTS

LÉVY FLIGHT OF HOLES IN InP SEMICONDUCTOR SCINTILLATOR

SERGE LURYI* and ARSEN SUBASHIEV†

Dept. of Electrical and Computer Engineering
State University of New York
Stony Brook, NY 11794-2350, USA
*serge.luryi@stonybrook.edu
†subashiev@ece.sunysb.edu

High radiative efficiency in moderately doped n-InP results in the transport of holes dominated by photon-assisted hopping, when radiative hole recombination at one spot produces a photon, whose interband absorption generates another hole, possibly far away. Due to "heavy tails" in the hop probability, this is a random walk with divergent diffusivity (process known as the Lévy flight). Our key evidence is derived from the ratio of transmitted and reflected luminescence spectra, measured in samples of different thicknesses. These experiments prove the non-exponential decay of the hole concentration from the initial photo-excitation spot. The power-law decay, characteristic of Lévy flights, is steep enough at short distances (steeper than an exponent) to fit the data for thin samples and slow enough at large distances to account for thick samples. The high radiative efficiency makes possible a semiconductor scintillator with efficient photon collection. It is rather unusual that the material is "opaque" at wavelengths of its own scintillation. Nevertheless, after repeated recycling most photons find their way to one of two photodiodes integrated on both sides of the semiconductor slab. We present an analytical model of photon collection in two-sided slab, which shows that the heavy tails of Lévy-flight transport lead to a high charge collection efficiency and hence high energy resolution. Finally, we discuss a possibility to increase the slab thickness while still quantifying the deposited energy and the interaction position within the slab. The idea is to use a layered semiconductor with photon-assisted collection of holes in narrow-bandgap layers spaced by distances far exceeding diffusion length. Holes collected in these radiative layers emit longwave radiation, to which the entire structure is transparent. Nearly-ideal calculated characteristics of a mm-thick layered scintillator can be scaled up to several centimeters.

Keywords: Photoluminescence; photon recycling; anomalous diffusion; scintillators.

1. Introduction

The term "Lévy flight" (LF) was coined by Benoît Mandelbrot to describe a random walk, in which step lengths ℓ have a probability distribution that is heavy-tailed. Although the exact definitions of "heavy tailing" vary in the literature, we shall reserve the term to distributions $\mathcal{P}(\ell)$ that do not possess a variance as they decrease too slowly at large steps, $\mathcal{P}(\ell) \propto \ell^{-(1+\gamma)}$. For the index γ in the range $0 < \gamma < 2$, the distribution itself can be normalized, $\int \mathcal{P}(\ell)\,d\ell = 1$, but its second moment, $\left\langle \ell^2 \right\rangle = \int \ell^2 \mathcal{P}(\ell)\,d\ell$, diverges. Although one often speaks of "anomalous diffusion", the LF random walk cannot be described by an ordinary diffusion equation. The conventional diffusivity is not even defined for such a random walk.

1

Possibility of statistical description of the random walk (e.g., through evaluation of a particle distribution that emerges from a point-type source after a given number of steps) relies on a statistical theorem that defines the limit of a sum of randomly distributed numbers (in our case these are the lengths of individual steps). If the step length distribution $\mathcal{P}(\ell)$ decreases rapidly enough for large steps (namely, when $\gamma > 2$) the result is given by the Central Limit Theorem (CLT) and the sum has a normal (Gaussian) distribution. When the steps are distributed with heavy tails, their sum does not follow the CLT and is not Gaussian. It may be still described by a universal (though γ-dependent) distribution, called the stable distribution. The first systematic studies of the stable distributions originate from Paul Lévy and Aleksandr Khinchin [1].

The LF transport problem has been extensively studied mathematically. Description of the anomalous transport in terms of fractional dynamic equations or, for random walks in external field, fractional Fokker-Planck equations, is amply discussed in the reviews [2-4]. These phenomena are well-known to astrophysicists, as they occur in the problem of transport of resonance radiation in celestial bodies [5, 6]. They are also known in plasma physics as the imprisonment of resonance radiation in gaseous discharge [7, 8]. Interestingly, LF transport is more common in nature than one might think: thus, Lévy flights were recently invoked to explain movement strategies in mussels as revealed in the patterning of mussel beds [9], as well as ocean predators search strategies in regions where prey is sparse [10]. Birds and other animals also seem to follow Lévy flights when foraging [11]. Finally, a vast literature is devoted to Lévy flights in finance, "random walk down the Street" [12].

Nevertheless, there have been preciously few experimentally available laboratory systems for studying LF transport, ideally with variable parameters. A rather ingenious such system was recently demonstrated by Barthelemy et al. [13], who embedded scattering particles in a glass matrix – together with non-scattering glass microspheres of same refractive index as the matrix. The sole purpose of these spacer spheres was to modify locally the average separation between the scattering particles and thus control the step-length distribution for photon transport. With specially designed, highly non-trivial, distributions of microspheres diameter, the authors were able to observe a Lévy flight of light.

Recently, we described [14] a more "natural" lab system exhibiting Lévy flight, namely the direct-gap semiconductor of high radiative efficiency, specifically n-doped InP. The randomly walking particles in this case are minority carriers (holes) and their dominant transport process is photon-assisted hopping. This process, also known as the photon recycling, consists of radiative recombination of a hole at one spot producing a photon, whose subsequent interband absorption leads to the re-emergence of a hole at another spot, possibly far away. The high radiative efficiency and low free-carrier absorption of light in lightly doped InP ensure that photon recycling continues for about 100 times before a hole recombines non-radiatively or a photon is absorbed without leaving a hole behind. The randomness of free flight is set by the emission spectrum in radiative recombination. This spectrum, combined with the interband absorption probability and the probability of photon propagation to a given distance, defines the probability distribution for free flights of photons. Photons generated in the long-wavelength wing of the emission spectrum travel long distances before they get re-absorbed and are responsible for the divergent variance of the distribution and the Lévy-flight nature of the resulting random walk. This process is reviewed in Section 2.

Manifestations of anomalous transport were found [14] by studying photo-luminescence in n-doped InP. The key evidence was derived from the ratio of transmitted and reflected luminescence spectra, measured in samples of the same doping level but very different thicknesses (350 µm vs. 50 µm). The results give a direct experimental proof of the non-exponential decay of the minority-carrier concentration from the surface where the holes were photo-excited initially. The power-law decay of the hole concentration, characteristic of the LF transport, is steep enough at short distances (steeper than an exponent) to fit the data for the thin sample, and at the same time slow enough at large distances (again, compared to an exponent) to account for the data for thick samples. This work is reviewed in Section 3.

Transport at much larger distances (up to centimeters) was studied in experiments [15], where photoluminescence was registered from the edge of an InP wafer as a function of the distance from the excitation spot on the broadside surface. Since the extremely long photon propagation is owing to the transparency region at the red wing of the emission spectrum, one observes a red shift in the luminescence spectrum, with larger shift corresponding to longer distances. Analysis of this shift provides an independent and accurate determination of the Urbach tails in moderately doped semiconductors. This work is reviewed in Section 4.

Sections 5 and 6 deal with practical applications of the anomalous transport of minority carriers in semiconductors of high radiative efficiency, specifically to the so-called semiconductor scintillator [16-18]. Normally, scintillators are not made of semiconductor material. The key issue in implementing a semiconductor scintillator is how to make the material transmit its own infrared luminescence, so that the response signal generated deep inside the semiconductor slab could reach its surface without tangible attenuation. In high-efficiency semiconductors, the long tails of Lévy-flight transport come to the rescue, providing near-ideal photon collection. Luminescence experiments [14, 15] support a simple model of photon collection, which we shall refer to as the "on the spot approximation" (OTSA). In this model, the signal received by a photodetector at the surface arises from repeated emission at the same spot where the initial minority carrier was generated. Each attempt has a small probability of "unhappy" termination, due to the nonradiative channel of recombination or free-carrier absorption. The happy end corresponds to the photon reaching the surface and being collected at the photodetector. The advantage of the OTSA is that it leads to a close-form expression for the collected signal, by summing a geometric series. As discussed in Section 5, the OTSA is very close to reality for the typical minority-carrier distributions generated by Lévy-flight transport.

Our understanding of the anomalous transport of minority carriers in direct-gap semiconductor of high radiative efficiency has led to the invention [18] of a layered scintillator, described in Section 6. The idea of embedding radiation sites in a semiconductor (or insulator) material is nearly as old as the scintillator concept itself [19]. In all such devices [20], the photo-generated carriers migrate to the radiation sites and recombine there emitting deep subband light, for which the material is transparent. The density of the radiation sites must be very high, so that the typical travel distance for carriers is much smaller than their diffusion length. The novelty of our idea [18] is to employ the photon-assisted transport of minority carriers rather than their ordinary diffusion. This allows one to space out the radiation sites (in our case, narrow low-bandgap wells embedded in a wide-gap semiconductor matrix) by a large distance. Ultimately, this may lead to the implementation of centimeter-thick semiconductor scintillators.

Our conclusions will be summarized in Section 7.

2. Photon Assisted Random Walk of Minority Carriers in InP

Suppose that an electron-hole pair is created by optical excitation in an n-doped infinite crystal. There is no interest in tracing the additional single electron as it produces little change in the majority-carrier system. All the action is due to the additional hole. Firstly (on a sub-picosecond time scale), it will become "thermalized", i.e. lose the excess energy it obtained from the light quantum. On a longer (nanosecond) time scale, the hole will move randomly with the thermal velocity until recombining with an electron. This type of random walk corresponds to the ordinary diffusion. The average hole lifetime τ depends on the electron concentration and is in the nanosecond range. The recombination process can be either radiative or non-radiative, and the rates of these processes are additive, $\tau^{-1} = \tau_{\rm rad}^{-1} + \tau_{\rm nr}^{-1}$. The probability of radiative recombination is described by the emission quantum efficiency η, viz.

$$\eta = \frac{\tau_{\rm nr}}{\tau_{\rm nr} + \tau_{\rm rad}} . \tag{2.1}$$

The non-radiative lifetime in high-quality crystals reaches several microseconds, resulting in $\eta > 90\%$. The emitted photons disappear mainly via interband absorption process, resulting in the generation of a new hole and then a new photon emitted via radiative recombination. The absorption-reemission sequence will be repeated many times until the recycled hole recombines non-radiatively or the intermediate photon is destroyed by a residual non-interband absorption process. This sequential process is called the photon recycling. Due to the short thermalization time of holes, the emission spectrum remains the same at all stages of the recycling and is well described by the equilibrium electron-hole recombination spectrum.

2.1. *Diffusion equation with a recycling term*

The spatial distribution of holes is formed by two additive transport processes. (*i*) the random flights of holes (at sub-micron distances) interrupted by scattering, as in the ordinary diffusion, and (*ii*) the photon-assisted transfer of holes over much larger distances. To quantify these processes we write down a modified diffusion equation for the concentration of holes $p(\mathbf{r},t)$:

$$\frac{\partial p}{\partial t} + D\Delta p = -\frac{p}{\tau} + G(\mathbf{r},t) + R(\mathbf{r},t), \tag{2.2}$$

where D is the diffusivity of holes, τ is the hole lifetime against all recombination processes, and $G(\mathbf{r},t)$ is the generation function defined as the concentration of holes generated per unit time. For a single hole generated at $\mathbf{r} = 0$ and $t = 0$, this function is $G(\mathbf{r},t) = \delta(\mathbf{r})\,\delta(t)$. The last term $R(\mathbf{r},t)$ is the recycling function,

$$R(\mathbf{r},t) = \frac{\eta}{\tau} \int P\big(|\mathbf{r} - \mathbf{r}'|\big)\, p(\mathbf{r}',t)\, d\mathbf{r}' , \tag{2.3}$$

which describes the concentration of holes generated per unit time at point $\mathbf{r}$ due to the radiative recombination of holes present in the crystal at the time t.

The factor $P\big(|\mathbf{r} - \mathbf{r}'|\big)$ in the integrand of Eq. (2.3) describes the probability that a hole at $\mathbf{r}'$ generates another hole at $\mathbf{r}$ by the described emission-reabsorption process. For the two points separated by the distance $r = |\mathbf{r} - \mathbf{r}'|$, this probability is given by

$$P(r) = \int \mathcal{N}(E)\, \frac{\exp[-\alpha_i(E)\,r]}{4\pi r^2}\, \alpha_i(E)\, dE, \tag{2.4}$$

where $\alpha_i(E)$ is the light absorption coefficient due to interband processes only. The integrand in Eq. (2.4) is the product of three probabilities: (*i*) the probability of emission of a photon of energy E, described by the normalized emission spectral function $\mathcal{N}(E)$; (*ii*) the propagation probability of this photon over the distance $r = |\mathbf{r} - \mathbf{r}'|$ (this probability is described by the intensity distribution produced by a unit point source) ; the absorption probability of this photon, described by the factor $\alpha_i(E)$. One can easily check that the probability (2.4) is properly normalized, $\int P(r)\,dV = 1$.

Equation (2.2) with the recycling term (2.3) was first obtained in the papers by Holstein [7] and Biberman [21] for the radiative spread of the excited-atom concentration in gases and is known as the Biberman-Holstein equation. Solution of this equation in 3D case is complicated by the fact that the resultant distribution cannot be factorized as a product of distributions along perpendicular axes. In other words, in contrast to the familiar Gaussian distribution, projections of the displacements on coordinate axes are correlated.[a] The source of these correlations and the entire difficulty reside in the recycling term. Nevertheless, one can find the solution of Eq. (2.2) quite generally by a trick due to V. A. Ambartsumyan [22], which transforms Eq. (2.2) for a point-like source into an equation for $\tilde{p}(z,t)$, which is the concentration $p(\mathbf{r},t)$ integrated over the (x,y) plane:

$$\tilde{p}(z,t) \equiv 2\pi \int_0^\infty p(z,\rho,t)\,\rho\,d\rho = 2\pi \int_z^\infty p(r,t)\,r\,dr \,, \qquad (2.5)$$

where $r^2 = z^2 + \rho^2$. With the known $\tilde{p}(z,t)$, one can find $p(r,t)$ by differentiating Eq. (2.5) with respect to z,

The Ambartsumyan transformation expresses the solution for a point-like source in 3D (the Green function of Eq. 2.2) in terms of the solution of a 1D problem with a plane source.

$$p(r,t) = -\frac{1}{2\pi r}\frac{\partial\,\tilde{p}(z,t)}{\partial z}\bigg|_{z=r}. \qquad (2.6)$$

The 1D concentration $\tilde{p}(z,t)$ obeys a much simpler equation that is obtained by integrating Eq. (2.2) over the (x,y) plane. The resulting 1D equation is of the form [23]

$$\frac{\partial\,\tilde{p}}{\partial t} - D\frac{\partial^2\,\tilde{p}}{\partial z^2} + \frac{\tilde{p}(z,t)}{\tau} = \frac{\eta}{2\tau}\int_{-\infty}^\infty \tilde{p}(z',t)\,\mathcal{P}\big(|z - z'|\big)\,dz' + \tilde{G}(z,t), \qquad (2.7)$$

where $\tilde{G}(z,t)$ is the generation term $G(\mathbf{r},t)$ integrated over the (x,y) plane. The probability $\mathcal{P}\big(|z - z'|\big)$ is given by

$$\mathcal{P}(z) = \int \mathcal{N}(E)\,\alpha_i(E)\,\mathrm{Ei}[1,\alpha(E)z]\,dE \,, \qquad (2.8)$$

---

[a] This can be illustrated in the instance when the characteristic function $F(\mathbf{k})$ of the distribution is of the form

$$F(\mathbf{k}) \equiv \int P(\mathbf{r})\exp[-i\mathbf{k}\cdot\mathbf{r}]dV = r_0|\mathbf{k}|$$

In the 1-dimensional case, the inverse Fourier transformation of $F(k_i)$ generates a Cauchy distribution,

$$P_1(x_i) = \frac{1}{\pi}\frac{r_0}{x_i^2 + r_0^2}$$

In the d-dimensional case, the transform of $F(\mathbf{k})$ yields

$$P_d(r) = \frac{\Gamma[(1+d)/2]}{\pi}\frac{r_0}{\left(r^2 + r_0^2\right)^{(1+d)/2}}\,, \text{ where } r^2 = \sum x_i^2 \,.$$

It is evident that the above expression for $P_d(r)$ cannot be factorized, and hence different x_i components are manifestly correlated.

where the exponential integral function is defined by

$$\mathrm{Ei}\,(1,z) = \int_1^\infty t^{-1} \exp(-zt)\,dt\,.$$

The probability $\mathcal{P}\big(|z - z'|\big)$ satisfies the normalization condition

$$\mathcal{P}_{\mathrm{tot}} \equiv \int_0^\infty \mathcal{P}(z)\,dz = 1\,. \tag{2.9}$$

If one knows $\mathcal{P}\big(|z - z'|\big)$, then, for an infinite medium, Eq. (2.7) can be solved [24] by a Fourier transformation. Using this equation, one can study the temporary evolution of the total number of holes per unit length along the z axis. In view of its linearity, Eq. (2.7) can be equally well applied to the case of a planar excitation uniform in the (x, y) plane – with $\tilde{G}(z,t)$ being the hole concentration generated per unit time. In this case, $\tilde{p}(z,t)$ is just the z-dependent concentration of holes.

In the problem of interest to us, the ordinary diffusion term gives negligible contribution. With this term dropped ($D = 0$), Eq. (2.7) retains a simple probabilistic interpretation: it describes a 1-dimensional random walk of a particle created at $z = 0$. The distribution of jump lengths is given by $\mathcal{P}\big(|z|\big)$, the average time between jumps is τ, and $1 - \eta$ is the probability of particle loss at any step. This interpretation suggests that Monte Carlo modeling should be a useful approach to studying $\tilde{p}(z,t)$. It has the advantage of being able to include various factors "difficult" in any analytic approach, such as effects of the boundaries, realistic shape of the generation pulse, etc. Our calculation begins with ascertaining the distribution $\mathcal{P}\big(|z|\big)$ of single jumps in the photon-assisted random walk of holes in n-InP.

2.2. *Jump distribution*

We shall use Eq. (2.8) to evaluate $\mathcal{P}\big(|z|\big)$ from the experimentally measured [15, 25] interband absorption coefficient $\alpha_i(E)$ for moderately doped n-type InP. With the known $\alpha_i(E)$, the spectral density $\mathcal{N}(E)$ of photon emission in the quasi-equilibrium hole recombination process (minority holes recombining with majority electrons) can be obtained by the thermodynamic relation due to van Roosbroek and Shockley [26],

$$\mathcal{N}(E) = A\,\alpha_i(E)\,E^2 e^{-E/kT}\,, \tag{2.10}$$

which we shall refer to as the VRS relation. Expression (2.10) represents the "intrinsic" emission spectrum and it agrees very well with the spectra of luminescence measured [27] from thin epitaxial layers (especially when those are clad by wider-gap layers to prevent surface recombination). Experimentally observed bulk luminescence spectra differ from VRS and the distortion depends on the geometry of the experiment. As discussed in Sections 3 and 4, the main contribution to spectral distortion results

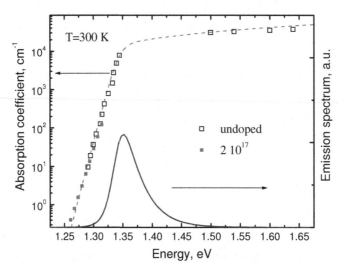

Fig. 2.1. Experimentally observed absorption spectrum (log scale) for moderately doped n-InP sample ($N_D = 2 \times 10^{17}$ cm^{-3}). Dashed lines show the fitting by Eq. (2.12). The intrinsic emission spectrum, derived from the VRS relation (2.10), is plotted on the linear scale.

from energy-dependent filtering due to the re-absorption of outgoing photons. These experimentally accessible filtering functions contain a wealth of information about the steady-state minority-carrier distribution $\tilde{p}(z)$. In this section, concerned with evaluation of $\mathcal{P}(|z|)$, we are not interested in filtering and assume that the intrinsic (unfiltered) emission lineshape is faithfully given by the VRS relation.

Both spectra, $\alpha_i(E)$ and $\mathcal{N}(E)$, are displayed in Fig. 2.1. Below the absorption edge, $\alpha_i(E)$ decreases exponentially

$$\ln \frac{\alpha_i(E)}{\alpha_i(E_G)} = \frac{E - E_G}{\Delta}. \tag{2.11}$$

If the bandgap E_G is independently known, the $\alpha_i(E)$ dependence is characterized by two parameters, absorption at the bandgap, $\alpha_i(E_G)$, and the tailing energy Δ (Urbach tail). In a broader range that includes the absorption edge and the region $E > E_G$ the dependence $\alpha_i(E)$ for moderately doped samples ($N_D < 10^{18}$ cm^{-3}) is well approximated by

$$\alpha_i(E) = \frac{E - E_0}{E_G} \frac{\alpha_0}{1 + \exp[(E_G - E)/\Delta]}, \tag{2.12}$$

see the dashed line in Fig. 2.1. Here the first factor reflects an almost linear growth of $\alpha_i(E)$ above the absorption edge and the second factor reproduces Urbach tailing (2.11). For the electron concentration $N_D = 2 \times 10^{17}$ cm^{-3}, parameters in (2.12) are $E_G = 1.35$ eV, $E_0 = 0.9$ eV, $\Delta = 8.2$ meV, and $\alpha_0 = 6.6 \times 10^5$ cm^{-1}. For higher concentrations, an increase is observed [15] in both E_G and Δ (for $N_D = 2 \times 10^{18}$ cm^{-3}, $E_G = 1.36$ eV, $E_0 = 1.2$ eV, $\Delta = 10.6$ meV, and $\alpha_0 = 1.2 \times 10^5$ cm^{-1}). Note that the Fermi level crosses E_G at $N_D = 4.3 \times 10^{17}$ cm^{-3} (at room temperature). At higher concentrations, the absorption spectra are influenced by the conduction band filling (the Moss-Burstein shift).

Numerical evaluation of $\mathcal{P}(|z|)$ with Eq. (2.8) is presented in Figs. 2.2 (a, b). In the entire range of z, the dependence is very close to

$$\mathcal{P}(z) = \frac{\gamma z_{min}^\gamma}{2 (z_{min} + z)^{1+\gamma}}, \tag{2.13}$$

where $z_{min} \approx 0.1$ μm is a distance scaling parameter that stems from truncation of the power law at short-distances and normalization of the distribution (2.13). The essential parameter is the exponent γ, called the index of the distribution. For moderately doped samples, in the range $N_D = (2 \text{ to } 6) \times 10^{17}$ cm^{-3} illustrated in Fig. 2.2 (a), the index varies from $\gamma \approx 0.69$ to 0.64, slightly decreasing with the doping level.

Theoretical calculation of the index γ in an analytical form can be carried out in a model, where the interband absorption is approximated by a function simpler than (2.12), viz.

$$\alpha_i(E) = \frac{\alpha_0}{1 + \exp[(E_G - E) / \Delta]}. \tag{2.14}$$

The function (2.14) decays exponentially below the absorption edge and saturates above it. This model correctly accounts for the Urbach tailing but it does not describe the approximately linear growth of $\alpha_i(E)$ at $E > E_G$. Furthermore, the emission spectrum in this model can be described by Eq. (2.10) with $\alpha_i(E)$ given by (2.14) and the pre-exponential factor replaced by its value at $E = E_G$ (reasonable for $\Delta \le kT << E_G$). The model yields a simple expression for the index,

$$\gamma = 1 - \frac{\Delta}{kT}. \tag{2.15}$$

Equation (2.15) predicts lower values of γ at lower temperatures. It also explains the decrease of γ with increasing N_D. The latter effect is due to the smearing of the absorption edge at higher doping, described by increasing tailing energy Δ. Estimation of the index with Eq. (2.15) for moderately doped samples ($\Delta = 9.4$ meV) gives $\gamma \approx 0.64$ in close agreement with the more accurate numerical calculations.

For heavier doping, the accurate numerical calculations yield a more complicated concentration dependence of γ (see Table I) compared to that predicted by (2.15). The discrepancy is due to the fact that Eq. (2.14) is no longer a good approximation when the Moss-Burstein shift is large.

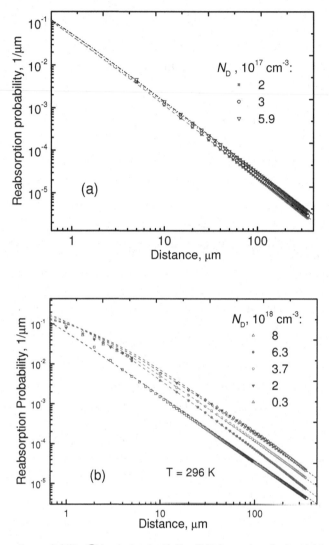

Fig. 2.2. Reabsorption probability $\mathcal{P}(z)$ calculated with Eq. (2.8) for moderately doped (a) and heavily doped (b) samples at room temperature. Relevant sample parameters are listed in Table I. Dashed lines correspond to the power-law approximation, Eq. (2.13). Note that the 3×10^{17} cm^{-3} sample is represented in both graphs for comparison.

Finally, we note that in addition to interband absorption, n-doped InP has a residual "free-carrier" absorption, $\alpha_{fc}(E)$, which linearly grows with the doping and weakly depends on the energy in the vicinity of the interband absorption edge. It can be easily taken into account by replacing $\alpha_i(E) \to \alpha(E) = \alpha_i(E) + \alpha_{fc}(E)$ in the exponential propagation probability factor in Eq. (2.4). In Eq. (2.8), similar replacement has been actually done tacitly in the argument of the Ei function. As a result, the full probability (2.9) of interband reabsorption becomes less than unity, $\mathcal{P}_{tot} < 1$. However, this effect is small. Experimentally, $\alpha_{fc} = 0.13 \times 10^{-17} N_D$ cm^{-1} at room temperature, and in the range of distances of our interest the effect of free-carrier absorption is negligible for all experiments discussed below.

Table I. Parameters of the jump distribution $\mathcal{P}(z)$ and the recycling factor Φ.

N_D (10^{18} cm^{-3})	0.2 to 0.6	2	3.7	6.3	8
γ	0.69 to 0.64	0.79	0.7	0.64	0.69
z_{min}(μm)	0.1	0.6	0.7	1.0	1.4
Φ	90	34	19	11	8

2.3. *Stable distribution of minority carriers*

The emission-reabsorption probability (2.8) describes a one-dimensional photon-assisted motion of holes that we call "jumps" in distinction from the actual hole movement interrupted by scattering and described as conventional diffusion. The stochastic nature of this 1D random walk is associated with the emission spectrum $\mathcal{N}(E)$. The hole jump probability, accurately described by Eq. (2.13), is typical for the "anomalous diffusion" of the Lévy-flight type [3]. Its hallmark is the asymptotic spatial decay with a "heavy tail," $\mathcal{P}(z) \propto 1/z^{1+\gamma}$, which has the power-law asymptotic index $\gamma < 2$. Since the second moment of this distribution diverges, one cannot describe the photon-assisted random walk of holes with a conventionally defined "enhanced" diffusion coefficient $D_{enh} \propto \langle z^2 \rangle / \tau$. However, for any finite number N of jumps z_i, one can find the statistical (averaged over many histories) hole distribution $p(z, N)$ for $z = \sum_{i=1}^{N} z_i$.

Monte Carlo simulation is well-suited for this purpose. The normalized hole distributions simulated with $\gamma = 0.76$ are presented in Fig. 2.3 (a) for several values of N. The initial excitation is localized at the origin. The heavy tails are very prominent on the logarithmic scale, especially in comparison with similar distributions shown in Fig. 2.3 (b) for the assumed index $\gamma > 2$, when $\langle z^2 \rangle$ is finite and the random walk is Brownian.

The statistical model of a random walk on an infinite line has been thoroughly studied [2, 3, 28]. After $N \gg 1$ jumps originating from $z = 0$, the distribution approaches the so-called *stable distribution* [28] with a given index $\gamma < 2$, viz.

$$p(z, N) = \frac{1}{\pi} \int_0^\infty \cos(kz) \exp\left[-N \left(z_C k \right)^\gamma \right] dk, \qquad (2.16)$$

where z_C is a depth scaling factor, which depends on the short distance behavior of the jump distribution $\mathcal{P}(z)$. In our case, $z_C \approx z_{min}$ by the order of magnitude. Comparing the distribution (2.16) with our Monte Carlo results, we find $z_C = 0.23$ μm. Figure 2.3 (a) displays both the results of Monte Carlo simulations and the calculation using Eq. (2.16). Excellent agreement demonstrates high accuracy of the Monte Carlo approach that will be extended below to include "difficult" factors, such as various boundary conditions in the random walk over a finite slab.

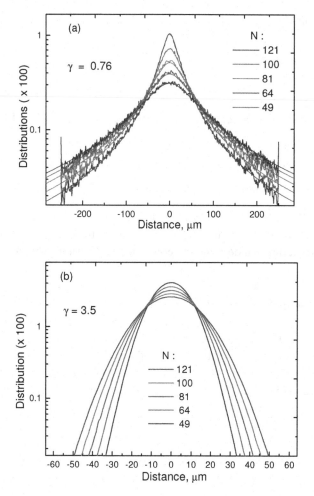

Fig. 2.3. (a) Hole distribution $p(z, N)$ calculated by Monte Carlo (noisy lines) assuming the jump probability $\mathcal{P}(z)$ in the form (2.13) with $\gamma = 0.76$ for an infinite crystal and holes generated at $z = 0$. Smooth lines are obtained by numerical evaluation of stable distribution (2.16) of same index.
(b) Similar results for $\gamma = 3.5$ and $z_{min} = 1$ μm. The displayed curves $p(z, N)$ calculated by Monte Carlo are very close to Gaussian curves of width $\langle z^2 \rangle = N \times (8/15) z_{min}^2$, evaluated according to the central limit theorem.

According to Eq. (2.16), the stable distribution is of the form

$$p(z, N) = \frac{1}{N^{1/\gamma}} g_\gamma \left(\frac{1}{N^{1/\gamma}} \frac{z}{z_C} \right), \tag{2.17}$$

with the universal function $g_\gamma(z)$ characterized by only one parameter γ. For this reason, the distribution is called "strictly stable", its universality being similar to the universally Gaussian shape of the normal distribution that emerges in the case of $\gamma > 2$, when the random walk is Brownian. Figure 2.3 (b) illustrates the Monte Carlo simulated distributions assuming $\mathcal{P}(z)$ of the form (2.13) with the index $\gamma = 3.5$. In this case, the 2nd moment is finite, $\langle z^2 \rangle = (8/15)\, z_{\min}^2$, and for large N, the distributions $p(z,N)$ are normal, according to the central limit theorem.

An important conclusion can be drawn by examining the asymptotic behavior of $g_\gamma(z)$. It follows from (2.16) that at large z

$$g_\gamma(z)\Big|_{z \gg z_C} \to \left(\frac{z_C}{z}\right)^{1+\gamma}, \tag{2.18}$$

which implies that

$$p(z,N)\,\big|_{z \gg z_C} \to N\left(\frac{z_C}{z}\right)^{1+\gamma}. \tag{2.19}$$

Comparison of Eqs. (2.13) and (2.19) illustrates a major property of the random walk with a heavy-tailed jump distribution: at large distances, the dominant contribution to $p(z,N)$ results from the single jumps from the starting point – with a pre-factor corresponding to the number of attempts. This result justifies the "on the spot approximation" (OTSA) that was mentioned in Introduction and will be put to practical use in Section 5.

2.4. *Stationary hole distribution for constant excitation*

In the preceding Section we calculated the distribution of holes $p(z,N)$ after a given number of jumps N upon their localized excitation, as if the hole jumps started all at once. We are now concerned with a *continuous* constant excitation and the resultant *stationary* distribution $p_{st}(z)$. To bridge these two problems, we consider the time evolution of the distribution $p(z,t)$ after a short excitation pulse at $t = 0$. Given the $p(z,t)$, we evaluate $p_{st}(z)$ by applying the Duhamel principle [29], viz.

$$p_{st}(z) = \int_0^\infty p(z,t)\exp(-t/\tau)\,dt/\tau, \tag{2.20}$$

where the exponential factor accounts for the decline of the hole concentration in time, with $\tau \approx \tau_{nr}$ being the average lifetime.[b]

---

[b] For the ordinary diffusion (Brownian random walk) one has $p(z,t) = (4\pi Dt)^{-1/2}\exp(-z^2/4Dt)$. In this case, Eq. (2.20) yields

$$p_{st}(z) = (2L_D)^{-1}\exp\left(-|z|/L_D\right),$$

which describes an exponential decline of the concentration away from the excitation point, with a characteristic length $L_D = \sqrt{D\tau}$, called the diffusion length.

Equation (2.20) suggest a Monte Carlo approach to evaluating $p_{st}(z)$. Firstly, we interpret the simulated distribution $p(z, N)$ as $p(z, t)$ for a fixed "discrete" time $t = N \tau_{rad}$ in units of the radiation emission time. Next, we average the simulated distributions over the durations of random walk t distributed as $\rho(t) = \tau_{nr}^{-1} \exp(-t / \tau_{nr})$. This is equivalent to averaging over the number of jumps (recycling events) N for a given mean value $\overline{N}$, which coincides by definition with the recycling factor $\Phi \equiv \overline{N}$. Thus, we have

$$p_{st}(z)\big|_{\Phi} = \sum_{N=1}^{\infty} p(z, N) \exp(-N / \Phi). \qquad (2.21)$$

For $\Phi \gg 1$, the main contribution to the sum for all $z > z_{min}$ comes from terms with $N \gg 1$ and hence both Eqs. (2.20) and (2.21) give very close results.

The stationary distribution for a given recycling factor Φ can also be found in an analytical form by substituting into (2.21) – instead of the Monte Carlo simulated distributions $p(z, N)$ – the quadrature expression (2.16) for the stable distribution. Substituting $N = t / \tau_{rad}$ in (2.16) and then integrating (2.20) with $\tau = \Phi \tau_{rad}$, we obtain

$$p_{st}(z)\big|_{\Phi} = \frac{1}{\pi} \int_{0}^{\infty} \frac{\cos(kz)\, dk}{1 + \Phi(kz_C)^{\gamma}}. \qquad (2.22)$$

Equation (2.22) is a new result that generalizes the stationary distribution for a Brownian ($\gamma > 2$) random walk to the case of a Lévy flight ($\gamma < 2$). Analysis of Eq. (2.22) readily shows the existence of two regions in the hole concentration profile, that of asymptotic decay and that of short jumps. In the asymptotic region the hole concentration is formed by repeated one-pass long-distance flights and hence it drops off like the jump probability itself, $p_{st}(z) \propto \mathcal{P}(z) \propto 1/z^{1+\gamma}$. The asymptotic region corresponds to $z \gg \Phi^{1/\gamma} z_C$. The length $z_F \equiv \Phi^{1/\gamma} z_C$ characterizes the front spread distance; in our samples $z_F \gg z_C$, since $\Phi \gg 1$ and $\gamma < 1$. In the short-jumps region, $z \ll z_F$, the concentration of holes drops off with distance at a much slower rate, $p_{st}(z) \propto 1/z^{1-\gamma}$.

Our Monte Carlo results for $p_{st}(z)$ are presented in Figs. 2.4 (a, b). We consider a relatively thin InP wafer with optical excitation near the front surface. The simulated distributions drop off with the distance in a non-exponential way, similar to Eq. (2.22) and characteristic of the LF transport. However, details are sensitive to the boundary conditions on the back surface. These effects are illustrated in Fig. 2.4 (a), which plots the $p_{st}(z)$ evaluated for samples with the same carrier concentration ($N_D = 6.3 \times 10^{18}$ cm^{-3}) but different boundary conditions. A particular strong effect is produced by the *reflecting back* boundary conditions. The corresponding stationary distribution decays with a much slower exponent compared to that for a semi-infinite medium (non-reflecting back surface). The reflecting boundary conditions are ruled out by our experiments (Section 3), even though on the first glance they appear plausible, owing to the complete internal reflection of most light rays at the back surface of InP wafer. However, these rays go outside the observation and are effectively extinct.

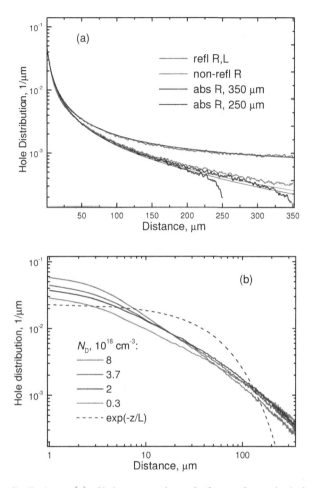

Fig. 2.4. Stationary distribution $p_{st}(z)$ of holes generated near the front surface and calculated by Monte Carlo assuming the jump probability $\mathcal{P}(z)$ in the form (2.13). Relevant distribution parameters for differently doped samples are listed in Table I.

(a) InP sample ($N_D = 6.3 \times 10^{18}$ cm^{-3}) of finite thickness and different boundary conditions at the back surface, **r**: reflecting, **a**: absorbing, **nr**: non-reflecting (the latter means a semi-infinite sample with no back surface). Dashed lines show the power-law fitting with $p_{st}(z) = c/(z + z_{min})^{1+\tilde{\gamma}}$ where $\tilde{\gamma} = 0.12$.

(b) Distributions $p_{st}(z)$ for differently doped samples with non-reflecting boundary conditions. The dashed line shows the exponential distribution for $L_D = 45$ μm (see footnote[b]).

The simulated distributions $p_{st}(z)$ for a set of differently doped samples and non-reflecting back boundary conditions are presented in Fig. 2.4 (b) on a log-log scale. Parameters of the jump probability $\mathcal{P}(z)$ and the recycling factor Φ are listed in Table I. For comparison, we also show an exponentially decaying distribution corresponding to a Brownian random walk with diffusion length $L_D = 45$ μm.

With the non-reflecting boundary conditions on the back surface, the simulated distributions are accurately described by Eq. (2.22). For moderately doped samples, one can estimate $z_F \approx 300$ µm and hence the onset of the asymptotic range is at distances larger than the sample thickness d. Neither the condition $d \ll z_F$ nor $d \gg z_F$ applies experimentally. In the sample range $0 < z < d$, the stationary distributions have an intermediate asymptotic $p_{st}(z) \propto 1/z^{1+\tilde{\gamma}}$ with $\tilde{\gamma} \approx 0.12 < \gamma$. The concentration extends over a much larger region than could be expected from an exponential distribution and it drops off slower than the single-jump probability $\mathcal{P}(z)$ given by Eqs. (2.8) and (2.13).

The normalized distributions sensitive to boundary conditions on the back surface, were found to be practically insensitive to the boundary conditions on the *front* surface.[c]

Finally, we remark that the one-dimensional Lévy-flight transport is fully described by the integro-differential equation (2.7). This equation, with appropriate boundary conditions, can be solved numerically, using available COMSOL software. We evaluated the hole distributions in this way and found excellent agreement with the Monte Carlo results, except within a region of the order of the hole diffusion length near the sample surface. Far away from that region, the integro-differential equation admits of an analytic solution [5], which is again Eq. (2.22).

3. Transmission and Reflection Luminescence Spectra

The basic experiment is illustrated in Fig. 3.1. Thin *n*-type InP wafers (mostly 350 µm, but some further thinned down to 250 µm and 50 µm) were illuminated by short wavelength radiation to ensure short penetration of the incident radiation into the wafer, so that the resulting distribution of holes is dominated by the carrier kinetics. The spectra were taken at room temperature in both reflection and transmission geometries. We shall denote these spectral functions by $I_{refl}(E)$ and $I_{trans}(E)$, and the total integrated luminescence intensities by L_{refl} and L_{trans}, respectively. The setup is schematically shown in Fig. 3.1 (a). It is convenient for the analysis to deal with the ratios of luminescence signals rather than the signals themselves, because the internal reflection factors at the wafer/air interface cancel out, being equal at the front and the back surface.

Figure 3.1 (b) shows the measured intensity ratio of the transmission and reflection luminescence as a function of the majority doping N_D. The ratio L_{trans}/L_{refl} increases both at high N_D (due to the Moss-Burstein effect) and at low N_D because of the enhanced photon recycling effect (higher recycling factor Φ).

The measured spectra $I_{refl}(E)$ and $I_{trans}(E)$ are distorted compared to the intrinsic emission spectrum $\mathcal{N}(E)$ in a different way, because of the different reabsorption-filtering geometry,

$$I_{trans}(E) = F_{trans}(E)\mathcal{N}(E), \tag{3.1a}$$

$$I_{refl}(E) = F_{refl}(E)\mathcal{N}(E). \tag{3.1b}$$

---

[c] For example, the stationary distributions $p_{rf}(z)$ obtained with the *reflecting front* boundary conditions satisfy for $z > 0$ the expected relation $p_{rf}(z) \approx 2p_\infty(z)$ with the distribution $p_\infty(z)$ corresponding to fully infinite media ($-\infty < z < \infty$). The factor of 2 corresponds to the contribution of an "image" source of photons provided by the reflecting boundary at $z = 0$.

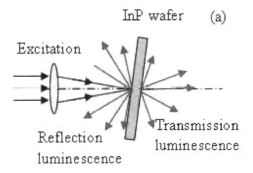

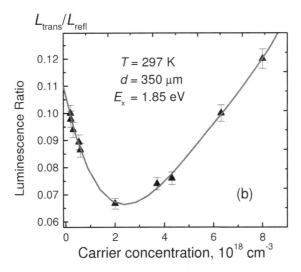

Fig. 3.1. Transmission and reflection luminescence.

(a) Schematic experimental setup. The excitation wavelength is chosen short enough to ensure small penetration of the incident radiation.

(b) Intensity ratio of the transmission and reflection luminescence as function of the doping N_{D}.

The filtering factors $F_{\mathrm{refl}}(E)$ and $F_{\mathrm{trans}}(E)$, are expressed through one-pass filtering functions,

$$F_1(E) = \int_0^d p(z)\, \exp[-\alpha(E)z]\, dz \, , \tag{3.2a}$$

$$F_2(E) = \int_0^d p(z)\, \exp[\alpha(E)(d-z)]\, dz \, . \tag{3.2b}$$

where $p(z)$ is the non-equilibrium stationary hole concentration that results from the steady-state excitation near the front surface $(z = 0)$ of the InP wafer of thickness d.

Taking into account multiple reflections at surfaces of the wafer, this expression is of the form [14],

$$F_{\text{trans}}(E) = (1 - R)\frac{F_2 + RF_1 \exp(-\alpha d)}{1 - R^2 \exp(-2\alpha d)}, \tag{3.3a}$$

$$F_{\text{refl}}(E) = (1 - R)\frac{F_1 + RF_2 \exp(-\alpha d)}{1 - R^2 \exp(-2\alpha d)}, \tag{3.3b}$$

where $R \approx 0.3$ is the InP reflection coefficient.[d]

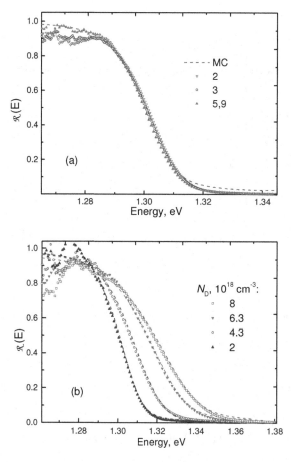

Fig. 3.2. Ratio $\mathcal{R}(E)$ of transmission and reflection luminescence spectra measured in 350 μm thick InP wafers of various doping levels N_D. For several select N_D we also show theoretical curves, calculated with Eqs. (3.4) and (3.2) with Monte Carlo modeled distributions $p(z)$ as in Fig. (2.4b).

(a) Lightly doped samples; (b) heavily doped samples.

---

[d] In numerical calculations [14], we took into account the measured dependence $R(E)$ and its variation with N_D.

It is possible *in principle* to experimentally determine both factors $F_{\text{refl}}(E)$ and $F_{\text{trans}}(E)$ from precisely measured spectra $I_{\text{refl}}(E)$, $I_{\text{trans}}(E)$ and $\alpha(E)$ [the latter determines $\mathcal{N}(E)$ by the VRS relation (2.10)]. Next, one could invert Eqs. (3.3) to determine functions $F_1(E)$ and $F_2(E)$, and use Eqs. (3.2) as integral equations for $p(z)$. In fact, viewed as functions of the argument $\alpha(E)$, the functions $F_1(\alpha)$ and $F_2(\alpha)$ represent Laplace transformations of $p(z)$, naturally extended to the infinite domain. Again *in principle*, one can obtain $p(z)$ by inverting these Laplace transformations numerically. However, numerical inversion of the Laplace transform is a classical ill-conditioned mathematical problem and the accuracy of our spectral measurements is not sufficient for finding a meaningful solution. Instead, our approach [14] was to calculate the reabsorption-filtering functions from the model distribution of holes in the layer, and compare the results with the functions found from the experiment.

The key experimental function for our analysis is the *ratio* of the transmission and reflection spectra, which in light of Eqs. (3.1) and (3.30) is given by

$$\mathcal{R}(E) \equiv \frac{I_{\text{trans}}(E)}{I_{\text{refl}}(E)} = \frac{F_2 + RF_1 \exp(-\alpha d)}{F_1 + RF_2 \exp(-\alpha d)} \ . \tag{3.4}$$

This ratio has important advantages for the analysis of the spatial hole distribution $p(z)$, firstly because it does not depend on details of the intrinsic emission spectrum. Furthermore, it is not sensitive to multiple reflections, since the denominators of Eq. (3.3) cancel out. At the same time, it is quite sensitive to $p(z)$ through $F_1(E)$ and $F_2(E)$. Therefore, the ratio $\mathcal{R}(E)$ is well suited to quantify the spatial hole distribution. Figure 3.2 shows the ratio for several samples, lightly doped (a) and heavily doped (b).

The experimental ratio curves show nearly perfect fit to the theoretical curves calculated from Eq. (3.4) with $F_1(E)$ and $F_2(E)$ given by Eq. (3.2), where we take for $p(z)$ the model stationary distributions $p_{\text{st}}(z)$ evaluated by Monte Carlo or by Eq. (2.22). The only adjustable parameter is the recycling factor, which was previously estimated independently from time-resolved luminescence kinetics [25]. The agreement is excellent.

Nevertheless, this fit, however perfect, does not provide an unambiguous evidence of Lévy flight transport. The "problem" is that we can obtain a reasonable fit also by assuming an exponential distribution (see footnote[b] above) with the length L_D as an adjustable parameter [14]. For example, the $N_\text{D} = 2 \times 10^{18}$ cm^{-3} data for $\mathcal{R}(E)$ are well fit with $L_\text{D} = 45\,\mu\text{m}$, cf. the exponential curve in Fig. 2.4 (b), corresponding to the Brownian random walk.

An unambiguous experimental demonstration of the LF transport is presented in Fig. 3.3, where we plot the ratio $\mathcal{R}(E)$ for samples of same doping but different thickness d. In this case, the theoretical fit based on distributions $p_{\text{st}}(z)$ evaluated for the Lévy flight remains nearly perfect, whereas the exponential distribution fails miserably. The exponential approximation fails to describe thin and thick samples simultaneously. To fit the data for 50 μm sample, one would have to assume length L_D to be substantially shorter than the sample thickness. This, however, would be in contradiction with the fairly high intensity of transmitted radiation in the 350 μm sample of the same doping. The power-law decay of the hole concentration is steep enough at short distances (steeper than an exponent) to fit the data for the thin sample, and at the same time slow enough at large distances (again, compared to an exponent) to account for thick samples.

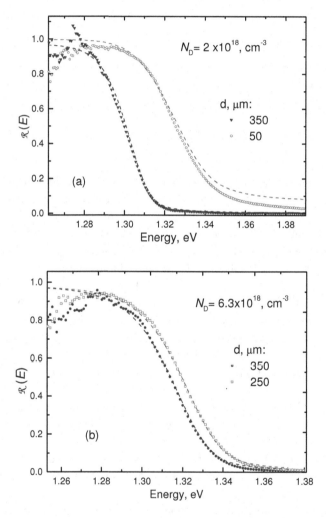

Fig. 3.3. Ratio $\mathcal{R}(E)$ of transmission and reflection luminescence spectra measured in a pair of thick and thin InP wafers of same doping concentration N_D. Dashed lines show theoretical curves, calculated with Eqs. (3.4) and (3.2) with Monte Carlo modeled model distributions $p(z) = p_{st}(z)$ as in Fig. (2.4b).

(a) Moderately doped samples, 350 μm and 50 μm thick;

(b) Heavily doped samples, 350 μm and 250 μm thick.

4. Luminescence Filtering and Urbach Tails

Luminescence studies described in Sect. 3 shed light on the anomalous transport properties over distances limited by the sample thickness, in our case < 350 µm. To circumvent this limitation, photoluminescence experiments were performed [15], where the luminescence spectra were excited by a red laser in a narrow spot on the broadside surface of an InP wafer but registered from the edge of the wafer, a distance d away from the excitation spot.

The observed spectra, Fig. 4.1, show two interesting features. Firstly, we see a power-law decrease of the integrated intensity and, secondly, a noticeable red shift of the spectral maximum. The power law is indicative of the anomalous transport but the exact Lévy-flight exponent is hard to extract from this experiment because of the irregular geometry producing specks of reflection. The red shift, on the other hand, can be analyzed very accurately and relate to the important spectral parameters, viz. the Urbach tails Δ and Δ' corresponding to absorption and emission spectra, respectively, and defined by their behavior deep in the red wing,

$$\alpha(E) = \alpha_0 \exp\left(\frac{E - E_G}{\Delta}\right), \tag{4.1a}$$

$$S(E) = S_0 \exp\left(\frac{E - E_G}{\Delta'}\right). \tag{4.1b}$$

Both tailing parameters depend on the doping concentration and the temperature, $\Delta = \Delta(T, N_D)$ and $\Delta' = \Delta'(T, N_D)$. In our $N_D = 3 \times 10^{17}$ cm^{-3} InP samples at room temperature, $\Delta = 9.4$ meV and $\Delta' = 15$ meV. For $N_D = 2 \times 10^{18}$ cm^{-3}, we have $\Delta = 10.6$ meV and $\Delta' = 16$ meV. The pre-exponential absorption factor α_0 cannot be measured independently of the small uncertainty in the bandgap E_G, but it should not depend on the doping and the value $\alpha_0 = 1.1 \times 10^4$ cm^{-1} fits very well in a wide temperature range for undoped samples, where E_G is known accurately.

The observed red shift, for several temperatures, fits very accurately to the expression,

$$E_{max}(d) = E_G - \Delta \ln\left(\frac{\alpha_0(d + d_{min})}{a}\right). \tag{4.2}$$

This expression comprises two empirical parameters, d_{min} and a. The former of these, reflects details of the experimental geometry (finite width and depth of the excitation spot) and, since $d_{min} < 200$ µm for all samples, it is of no importance when distances d are in the range of 1 to 20 mm, i.e., for $d \gg d_{min}$. The main empirical content of the dependence $E_{max}(d)$ resides in the parameter $a = a(T, N_D)$. For the samples with $N_D = 3 \times 10^{17}$ cm^{-3} and $N_D = 2 \times 10^{18}$ cm^{-3} at $T = 300$ K, we have respectively $a = 0.63$ and $a = 0.68$.

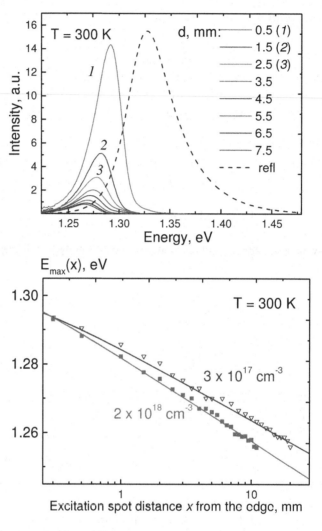

Fig. 4.1. Edge luminescence experiment [15].

(a) Room-temperature luminescence spectra for n-InP sample, $(3 \times 10^{17}\ \mathrm{cm}^{-3})$ observed at increasing distances d between the excitation spot and the wafer edge; the dashed line shows the emission spectrum;

(b) Shift of the luminescence peak $E_{max}(d)$ for two samples of different doping.

The observed $E_{max}(d)$, including the values of a, can be reproduced in a simple model that attributes the luminescent peak shift to wavelength-dependent filtering of outgoing radiation by the sample absorption. In the spirit of the on-the-spot approximation, the observed radiation at the edge arises from repeated emissions at the same spot where the initial hole was generated. Therefore, we assume that the position of the peak observed at distance d from the excitation spot is determined by the transparency of InP wafer to the emission spectrum $S(E)$. In other words, the observed edge spectrum near its maximum is described by

$$S_{\text{obs}}(E,d) = S(E) \times \exp\left[-\alpha(E)d\right]. \tag{4.3}$$

The strong refraction of outgoing radiation and a relatively small observation angle ensure a small and constant range of the angles of incidence. Therefore, the d dependence corresponds to one-dimensional attenuation of light. The maximum of the observed spectrum can be found from the expression $dS_{\text{obs}}(E,d) / dE = 0$, which, in light of (4.3), takes the form

$$\left.\frac{d\ln[S(E)]}{dE}\right|_{\text{max}} = d \times \left.\frac{d\alpha(E)}{dE}\right|_{\text{max}}, \tag{4.4}$$

Substituting Eqs. (4.1) into (4.4), we find an expression of the form (4.2) with the parameter a given by $a = \Delta / \Delta'$. For $N_{\text{D}} = 3 \times 10^{17}$ cm^{-3} and $N_{\text{D}} = 2 \times 10^{18}$ cm^{-3}, respectively, the Urbach tail ratio gives $a = 0.63$ and $a = 0.67$ in a remarkable agreement with the above empirical values. Similar agreement is obtained for other samples at all temperatures [15].

For all studied cases, the values of Δ obtained from the slope of $\ln \alpha(E)$ and the slope of the dependence of $E_{\text{max}}(\ln d)$ are very close, the difference never exceeding 0.2 meV. Thus, the described edge luminescence method provides an independent way of measuring the tailing parameters. This method can be indispensable (in fact, the only available) in the case when the residual absorption is strong.

Edge luminescence studies lend further support to the on-the-spot approximation (OTSA) that was justified theoretically in Section 2: at large distances, the dominant contribution to the observed spectra results from repeated jumps from the starting point. This principle will be used in the evaluation of photon collection efficiency in semiconductor scintillators, Sections 5 and 6.

5. Photon Collection Efficiency in InP Scintillator

The key issue in implementing a semiconductor scintillator is to make sure that photons generated deep inside the semiconductor slab could reach its surface without tangible attenuation. However, semiconductors are usually opaque at wavelengths corresponding to their radiative emission spectrum. Our group has been working on the implementation of scintillators based on direct-gap semiconductors. For the exemplary case of InP, the luminescence spectrum is a band of wavelengths near 920 nm. The original idea [30] was to make InP relatively transparent to this radiation by doping it heavily with donor impurities, so as to introduce the Burstein shift between the emission and the absorption spectra.

Here we shall describe another approach [16], based on the photon recycling effect owing to the high radiative efficiency of best direct-gap semiconductors, such as InP. In these materials, an act of interband absorption does not finish off the luminescent photon; it merely creates a new minority carrier and then a new photon in a random direction. The resultant random walk has been described (Section 2) as the Lévy flight of holes (or photons).

Consider an InP scintillator slab with two photoreceiver systems integrated on the opposite sides of the slab [31]. Exemplarily, these are epitaxially grown InGaAsP photodiodes [32]. Let the interaction occur a distance z from the detector top surface, as indicated in Fig. 5.1, producing minority carriers (holes). A hole has the probability η (the radiative efficiency, Eq. 2.1) to generate a photon (distributed in energy according to the emission spectrum, Eq. 2.10). The generated photon can either reach the detectors (probabilities π_1 and π_2, respectively) or disappear through free-carrier absorption (single-pass probability π_{FCA}). All these probabilities depend on z. The combined probability $\Pi(z) = \pi_1 + \pi_2 + \pi_{\mathrm{FCA}}$ describes the likelihood of the photon loss at this stage and the alternative, $1 - \Pi(z)$, is the probability that a new hole is created. The cycle of hole-photon-hole transformation repeats *ad infinitum*. Most of the scintillation reaching the detectors' surface are not photons directly generated at the site of the gamma particle interaction, but photons that have been re-absorbed and re-emitted a multiple number of times.

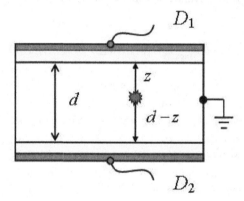

Fig. 5.1. Schematic cross-section of InP scintillator with two epitaxial photodiodes grown on both sides. Interaction with incident gamma photon (shown by the explosion symbol) occurs a distance z from the top surface and both photodiode signals depend on this distance.

The detector signals D_1 and D_2 add single-pass contributions from different cycles. The sum can be found as geometric progression, giving (per unit strength of initial excitation)

$$D_i(z) = \eta \pi_i(z) \times \sum_{n=0}^{\infty} \left[\eta(1-\Pi) \right]^n = \frac{\eta \pi_i(z)}{(1-\eta) + \eta \Pi(z)}, \qquad (5.1)$$

where $i = 1, 2$. Taking into account Eq. (2.1), the photon collection efficiency, PCE $\equiv D_1 + D_2$ is given by

$$\mathrm{PCE} = \frac{\pi_1(z) + \pi_2(z)}{[(\tau_{\mathrm{rad}} / \tau_{\mathrm{nr}}) + \pi_{\mathrm{FCA}}(z)] + [\pi_1(z) + \pi_2(z)]}. \qquad (5.2)$$

We note that for high photon recycling ($\eta \rightarrow 1$ and $\pi_{\text{FCA}} \rightarrow 0$), the entire luminescence is collected – even though the single-pass probabilities π_1 and π_2 may not be high due to interband absorption. The efficiency of photon collection is thus limited by parasitic processes, such as FCA and nonradiative recombination of holes. If these are minimized, one can have an "opaque" but ideal semiconductor scintillator.

The only approximation involved in Eq. (5.1) is the assumption that every act of recycling occurs at the same place z where the initial interaction occurred, and therefore the same probabilities $\pi_1(z)$ and $\pi_2(z)$ appear at all stages of the recycling. This "on-the-spot" approximation (OTSA) has reduced the summation of an infinite series to a geometric progression and allowed us to obtain the result in a closed form. The physical motivation for OTSA (the Lévy flight nature of random walk involved in the recycling) was presented in Sects. 2c and 4. Here we remark that going beyond OTSA (by direct Monte Carlo evaluation of PCE) does not change the results qualitatively, although it slightly enhances our estimate of PCE.

The single-pass probabilities $\pi_1(z)$ and $\pi_2(z) = \pi_1(d - z)$ can be evaluated by integration over the isotropic distribution of photon directions and the random energies in the emission spectrum,

$$\pi_1(z) = \int \mathcal{N}(E)\,\pi(z, E)\,dE\ ,$$

$$\pi(z, E) = \int_0^\infty \exp[\alpha_i(E)r]\,\frac{\cos\theta}{2r^2}\,\rho d\rho\ , \tag{5.3}$$

where $\rho = z\tan\theta$ and $r = z\sec\theta$. Similarly, we can evaluate the single-pass probability of free-carrier absorption in terms of the FCA coefficient α_{fc}. If we neglect in the exact expression [16] for $\pi_{\text{FCA}}(z)$ the corrections due to $\pi_1(z)$ and $\pi_2(z)$, but retain the dominant process of interband absorption α_i, then π_{FCA} no longer depends on z, viz.

$$\pi_{\text{FCA}} = \int \mathcal{N}(E)\,\frac{\alpha_{fc}(E)}{\alpha_{fc}(E) + \alpha_i(E)}\,dE\ . \tag{5.4}$$

Practically, for moderately doped samples, one has $\pi_{\text{FCA}} \ll \tau_{\text{rad}}/\tau_{\text{nr}}$ and FCA can be neglected. The calculated PCE is shown in Fig. 5.2 (a) for three moderate doping concentrations. We see that photon recycling delivers a reasonable fraction of the scintillation to the wafer surface, this fraction being higher for samples with higher radiative efficiency. We note, however, that the photon yield depends on the exact position of the interaction relative to the wafer surfaces. This spells trouble for the needed precise quantification of the energy deposited by a gamma quantum. The problem is how to distinguish the signal arising from a large energy deposited far from the photoreceiver surface from that arising from smaller energy deposited nearby. The problem arises from the attenuation of the optical signal.

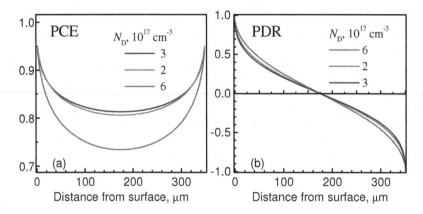

Fig. 5.2. Model calculations in the "on-the-spot" approximation for 350 μm thick InP wafers of different doping concentration at room temperature.

(a) Photon collection efficiency PCE (z) calculated according to Eq. (5.2) as a function of the position z of the interaction

(b) Position determining ratio PDR (z) calculated from Eq. (5.5) as a function of z.

However, if we knew the position z of the gamma interaction event, we could correct for the attenuation. The solution [31] is based on tallying the signals D_1 and D_2 individually. The relative strength of the two signals provides a good measure of event position. A convenient characteristic is the position determining ratio, defined by PDR $\equiv (D_1 - D_2)/(D_1 + D_2)$ and plotted in Fig. 5.2 (b) as a function of position, PDR (z). We see that the PDR is an excellent measure of z.

From Eq. (5.1) we find

$$\text{PDR} = \frac{\pi_1(z) - \pi_2(z)}{\pi_1(z) + \pi_2(z)}. \tag{5.5}$$

The simultaneous detection by *both* detectors of the scintillation arising from the same interaction event, allows us to determine the position of the event and correct for attenuation.

As seen from Eq. (5.2), the efficiency of photon collection in a scintillator based on photon recycling depends strongly on the radiative efficiency of the material and is maximized when $\eta \rightarrow 1$, or, equivalently, when $\tau_{rad} / \tau_{nr} \rightarrow 0$. Since $1/\tau_{rad} = BN_D$ and $1/\tau_{nr} = A + CN_D^2$, this ratio is non-monotonic in concentration [17] and has a minimum when $N_D^2 = A/C$. In our series of InP samples, the optimum doping is $N_D = 3 \times 10^{17}$ cm^{-3}, where $\eta \approx 98\%$ or even higher [14, 17]. Unfortunately, the high radiative efficiency of the low-doped InP scintillator material does not survive the high-temperature treatment involved in the epitaxial growth and processing of the quaternary InGaAsP *pin* diodes [32]. The bright luminescence of the virgin wafers degrades by nearly two orders of the magnitude upon the heat treatment. The degradation is thermally activated and is apparently related to defects inherent in a bulk Czochralski-grown wafer.

We have been therefore led to explore the possibility of all-epitaxial scintillator. It is known that the luminescent properties of low-doped epitaxial layers, as opposed to those of a bulk wafer, *do not* degrade under high-temperature treatment. Thick (e.g., millimeter-thick) free-standing layers can be grown by such epitaxial techniques as HVPE ("Hydride Vapor Phase Epitaxy," a growth technique with rates exceeding 100 μm an hour [33]) and can be expected to have superior non-degrading luminescence properties. While we began to pursue such an approach experimentally, we explored theoretically the possibilities it offers. The most recent advance in this regard was our invention [18] of an artificially layered scintillator material that comprises alternating thick wide-gap "barrier" and thin narrow-gap "well" layers. The wells constitute the radiation sites that are *pumped by light* generated in the barriers. The idea is discussed in the next Section.

6. Layered Scintillator Based on Photon-Assisted Transport of Holes to Radiation Sites

Semiconductor scintillator structure that could be scaled to thicknesses $d \approx 1$ mm and higher would be very attractive. We decided to explore the possibilities that would arise with an all-epitaxial fast-growth technique. This has led to our invention [18] described below.

The inventive scintillator material illustrated in Fig. 6.1 is a direct bandgap semiconductor heterostructure that comprises alternating thick wide-gap "barrier" layers B_j and thin narrow-gap "well" layers W_j. Exemplarily, we take the wide-gap layers as low-doped InP and the well layers as lattice-matched InGaAsP alloy of 100 meV narrower bandgap. The assumed absorption and emission spectra shown in Fig. 6.2 are based on the experimental spectra of Fig. 2.1.

Minority carriers generated in the wide-gap material by an incident high-energy particle, recombine there radiatively, producing *primary* scintillation light that is captured by the narrow-gap wells generating new minority carriers therein. Recombination of these new minority carriers in the narrow-gap wells generates *secondary* longer wavelength scintillation – to which the entire layered structure is largely transparent. It is important that the separation between narrow-gap wells *is not limited by the minority-carrier diffusion length* and can be as large as hundreds of microns.

In the conventional scintillator language, the narrow-gap wells can be viewed as radiation sites that emit light at subband wavelengths [30, 34]. However, no travel of carriers to these radiation sites is now contemplated. This is a radical departure from all prior-art activated scintillators, where charge carriers were supposed to travel to the radiation sites and the distance to travel had to be minimized by increasing their concentration. The finite travel distance, even when minimized, leads to an unwelcome "non-proportionality" [35] of activated dielectric scintillators, which impedes their energy resolution. The present invention circumvents this requirement.

The main advantage of the inventive scintillator structure is the possibility to enhance the overall thickness of the semiconductor body beyond 1 mm.

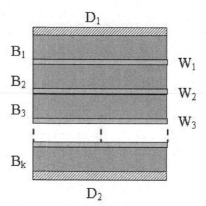

Fig. 6.1. All-epitaxial multilayered scintillator [18]. The scintillator body comprises a sequence of k barrier (InP) layers B_i of thickness b, alternating with $k-1$ well layers W_i of thickness w made of quaternary InGaAsP lattice-matched to InP. The wells collect the primary photons generated in the barriers and in turn generate secondary photons – the scintillation output. The structure includes photoreceivers D_1 (top) and D_2 (bottom), sensitive to the scintillation. Exemplarily, $k = 10$, $b = 100$ μm and $w = 1$ μm, so there are altogether 10 barriers and 9 wells of total thickness about 1 mm.

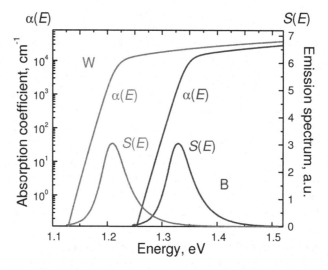

Fig. 6.2. Absorption and emission spectra of the multilayer structure, assumed in our calculations. For barrier layers B, the assumed spectra coincide with experimental curves (Fig. 2.1) for moderately doped InP. For the well layers W, the assumed curves describe similarly-doped lattice-matched InGaAsP alloy whose bandgap is 100 meV narrower than that of InP.

Figure 6.3 shows the results of calculating the response of the 1 mm thick multilayered structure described by Figs. 6.1 and 6.2. Calculations presented are based on the absorption and emission spectra of InP and take full account of the anomalous transport properties of photons in the high-radiative efficiency InP material.

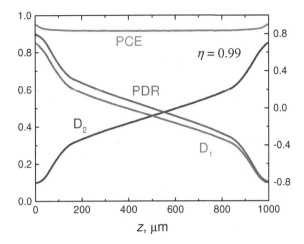

Fig. 6.3. Normalized photodiode signals D_1 and D_2, the photon collection efficiency (PCE = $D_1 + D_2$), and the position-determining ratio PDR = $(D_1 - D_2)/(D_1 + D_2)$ calculated for the multilayered structure described by Figs. 6.1 and 6.2 as a function of the position z of the interaction, counted from the bottom photoreceiver plane. The assumed quantum radiative efficiency is η = 99 % throughout the structure.

Photon-assisted transport enables us to space the radiation sites by a large distance, much larger than the diffusion length of carriers. Ultimately, this may lead to the implementation of centimeter-thick semiconductor scintillators. The capability for vertical position determination by the PDR function makes thick scintillators very attractive for their desired application [17] as a voxel (3D pixel) in a three-dimensional array of radiation detectors. The PDR resolves the z coordinate to a much finer degree than the linear dimension of the voxel in the z direction and enables one to replace a 3D stack of 2D arrays by a single 2D array of thick layered scintillators.

7. Conclusion

We have discussed a remarkable laboratory system for studying Lévy flights, namely high-radiative-efficiency semiconductors. The Lévy-flight model provides an adequate description for the photon-assisted transport of minority carriers, where photons are repeatedly recycled by the interband absorption/re-emission processes. Based on photoluminescence experiments in high-efficiency bulk InP, we have demonstrated an unambiguous evidence for the Lévy-flight nature of this anomalous non-diffusive transport.

We have developed a quantitative theoretical description of the steady-state distributions of minority carriers that emerge in the anomalous transport. The description, based on the mathematics of the *stable* Lévy-flight distributions, has been checked against both the experiments and Monte Carlo simulations. Our key experiments involve spectral *ratios* of the luminescence observed in the transmission and reflection geometries. Particularly revealing are ratios obtained with the variable wafer thickness.

Supported by the Lévy-flight model, we have formulated the so-called *on-the-spot approximation* (OTSA) that proves to be very useful in describing photon collection processes in semiconductor scintillators based on photon recycling. Equations of the OTSA demonstrate the tantalizing possibility of implementing a direct-gap semiconductor scintillator that is *opaque* in the usual sense at the wavelength of its own scintillation. Nevertheless, this scintillator will have nearly ideal photon collection efficiency.

Finally, we discussed our recent invention, not yet reduced to practice, of a *layered* semiconductor scintillator, in which the Lévy-flight photon-assisted transport is used to deliver the generated primary minority carriers to the "radiation sites" implemented as semiconductor wells of narrower bandgap, spaced apart by distances much larger than the diffusion length of the minority carriers. The nearly ideal characteristics of the layered millimeter-thick scintillator can be scaled up to the dimensions of several centimeters.

Acknowledgements

This work was supported by the Domestic Nuclear Detection Office (DNDO) of the Department of Homeland Security, by the Defense Threat Reduction Agency (DTRA) through its basic research program, and by the New York State Office of Science, Technology and Academic Research (NYSTAR) through the Center for Advanced Sensor Technology (Sensor CAT) at Stony Brook.

References

1. P. Lévy, *Calcul des probabilités*, Paris, Gautier-Villars (1925); A. Khinchine and P. Lévy, "Sur les lois stables", *Comptes Rendus de l'Académie des Sciences* **202**, pp. 274-276 (1936).
2. R. Metzler and J. Klafter, "The restaurant at the end of the random walk: recent developments in the description of anomalous transport by fractional dynamics," *J. Phys. A*: *Math. Gen.* **37**, pp. R161-R208 (2004).
3. A. V. Chechkin, V. Yu. Gonchar, J. Klafter, and R. Metzler, "Fundamentals of Lévy Flight Processes," *Advances in Chemical Physics* **133**, Part B, Chapter 9, pp. 439-496 (2006).
4. R. Metzler, A. V. Chechkin, V. Yu. Gonchar, and J. Klafter, "Some fundamental aspects of Lévy flights," *Chaos, Solitons & Fractals* **34**, pp. 129-142 (2007).
5. V. V. Ivanov, *Transfer of radiation in spectral lines* (Revised version of V. V. Ivanov's *Radiative Transfer and Spectra of Celestial Bodies*, in Russian, Moscow, 1969), National Bureau of Standards Special Publication **385** (1973).
6. G. B. Rybicki and A. P. Lightman, *Radiative Processes in Astrophysics*, Wiley, NY (1979).
7. T. Holstein, "Imprisonment of resonance radiation in gases," *Phys. Rev.* **72**, pp. 1212-1233 (1947); "Imprisonment of resonance radiation in gases. II," *ibid* **83**, pp. 1159-1168 (1951).
8. E. Pereira, J. M. G. Martinho and M. N. Berberan-Santos, "Photon Trajectories in Incoherent Atomic Radiation Trapping as Lévy Flights," *Phys. Rev. Lett.* **93**, 120201:1-4 (2004).
9. M. de Jager, F. J. Weissing, P. M. J. Herman, B. A. Nolet, and J. van de Koppel, "Lévy Walks Evolve Through Interaction Between Movement and Environmental Complexity," *Science* **332**, pp. 1551-1553 (2011); Errata: *ibid.* **334**, p. 1641 (2011).
10. N. E. Humphries, N. Queiroz, J. R. M. Dyer, N. G. Pade, M. K. Musyl, K. M. Schaefer, D. W. Fuller, J. M. Brunnschweiler, T. K. Doyle, J. D. R. Houghton, G. C. Hays, C. S. Jones, L. R. Noble, V. J. Wearmouth, E. J. Southall, and D. W. Sims, "Environmental context explains

Lévy and Brownian movement patterns of marine predators," *Nature* **465**, pp. 1066-1069 (2010).

11. G. M. Viswanathan, S. V. Buldyrev, S. Havlin, M. G. E. da Luz, E. P. Raposo, and H. E. Stanley, "Optimizing the success of random searches," *Nature* **401**, pp. 911-915 (1999).

12. Benoît Mandelbrot, *Fractals and Scaling in Finance*, Springer, NY (1997).

13. P. Barthelemy, J. Bertolotti, and D. S. Wiersma, "A Lévy flight for light," *Nature* **453**, pp. 495-498 (2008).

14. O. Semyonov, A. V. Subashiev, Z. Chen, and S. Luryi, "Photon assisted Lévy flights of minority carriers in *n*-InP," *Journal of Luminescence* **132**, pp. 1935-1943 (2012).

15. A. Subashiev, O. Semyonov, Z. Chen, and S. Luryi, "Urbach tail studies by luminescence filtering in moderately doped bulk InP," *Appl. Phys. Lett.* **97**, 181914 (2010).

16. S. Luryi and A. Subashiev, "Semiconductor scintillator based on photon recycling," *Nuclear Inst. and Methods in Physics Research A* **652**, pp. 292-294 (2011).

17. S. Luryi and A. Subashiev, "Semiconductor Scintillator for 3-Dimensional Array of Radiation Detectors," in *Future Trends in Microelectronics: From Nanophotonics to Sensors to Energy*, ed. by S. Luryi, J. M. Xu and A. Zaslavsky, Wiley, Hoboken, NJ (2010) pp. 331-346.

18. S. Luryi and A. Subashiev, "Layered Semiconductor Scintillator," US Patent application, No. **13/316,706** (filed Dec 2011).

19. R. Hofstadter, "Alkali Halide Scintillation Counters," *Phys. Rev.* **74**, pp. 100-101 (1948).

20. G. F. Knoll, *Radiation Detection and Measurement*, 3rd ed., Wiley, NY (2000).

21. L. M. Biberman, "On the diffusion theory of resonance radiation," *Zh. Eksp. Teor. Fiz.* **17**, 416 (1947) [Eng. Transl.: *Sov. Phys. – JETP* **19**, pp. 584-603 (1949)].

22. V. A. Ambartsumyan ["On the Scattering of Light by a Diffuse Medium," *Comptes Rendus (Doklady) de l'Académie des Sciences U.R.S.S.* **38**, pp. 257-261 (1943)] considered a more general problem with the radiation not only being reabsorbed but also elastically scattered. His approach was further developed in the context of astrophysics, as discussed in Ref. [5], Chap. 3, and in the book by S. Chandrasekhar, *Radiative transfer*, Dover, NY (1960) Chap. 4.

23. V. V. Rossin and V. G. Sidorov, "Reabsorption of recombination radiation in semiconductors with high internal quantum efficiency," *Phys. Stat. Solidi A* **95**, pp. 15-40 (1986).

24. B. A. Veklenko, "On the Green function for the resonance radiation diffusion equation," *Zh. Eksp. Teor. Fiz.* **36**, pp. 204-211 (1959) [Eng. Transl.: *Sov. Phys. – JETP* **9**, pp. 138-144 (1959)].

25. O. Semyonov, A. V. Subashiev, Z. Chen, and S. Luryi, "Radiation efficiency of heavily doped bulk n-InP semiconductor," *J. Appl. Phys.* **108**, 013101, pp. 1-7 (2010).

26. W. van Roosbroek and W. Shockley, "Photon-Radiative Recombination of Electrons and Holes in Germanium", *Phys. Rev.* **94**, 1558 (1954); the VRS relation is sometimes also referred to as the Kubo-Martin-Schwinger theorem, cf. R. Kubo, "Statistical-Mechanical Theory of Irreversible Processes. I. General Theory and Simple Applications to Magnetic and Conduction Problems," *J. Phys. Soc. Jpn.* **12**, pp. 570-586 (1957); P. C. Martin and J. Schwinger, "Theory of Many-Particle Systems (I)," *Phys. Rev.* **115**, pp. 1342-1349 (1959).

27. R. M. Sieg and S. A. Ringel, "Reabsorption, band-gap narrowing, and the reconciliation of photoluminescence spectra with electrical measurements for epitaxial n-InP," *J. Appl. Phys.* **80**, pp. 448-458 (1996).

28. V. V. Uchaikin and V. M. Zolotarev, *Chance and Stability: Stable Distributions and Their Applications*, VSP, Utrecht (1999).

29. M. Necati Özişik, *Boundary Value Problems of Heat Conduction*, Dover, NY (1989) Ch. 5.

30. A. Kastalsky, S. Luryi, and B. Spivak, "Semiconductor high-energy radiation scintillation detector," *Nucl. Instr. and Meth. in Phys. Research A* **565**, pp. 650-656 (2006).

31. J. H. Abeles and S. Luryi, "Slab scintillator with integrated two-sided photoreceiver," US Patent **8,253,109** (issued Aug 2012).

32. S. Luryi, A. Kastalsky, M. Gouzman, N. Lifshitz, O. Semyonov, M. Stanacevic, A. V. Subashiev, V. Kuzminsky, W. Cheng, V. Smagin, Z. Chen, J. H. Abeles, W. K. Chan, and Z. A. Shellenbarger, "Epitaxial InGaAsP/InP photodiode for registration of InP scintillation," *Nucl. Instr. and Meth. in Phys. Research A* **622**, pp. 113-119 (2010).
33. S. Lourdudoss and O. Kjebon, "Hydride Vapor Phase Epitaxy Revisited," *IEEE J. Selected Topics Quant. Electronics* **3**, pp. 749-767 (1997).
34. Serge Luryi, "Impregnated Semiconductor Scintillator," *International Journal of High Speed Electronics and Systems* **18**, pp. 973-982 (2008).
35. A. N. Vasil'ev, "From luminescence non-linearity to scintillation non-proportionality," *IEEE Trans. Nucl. Sci.* **55**, pp. 1054-1061 (2008).

InAs$_{1-x}$Sb$_x$ ALLOYS WITH NATIVE LATTICE PARAMETERS GROWN ON COMPOSITIONALLY GRADED BUFFERS: STRUCTURAL AND OPTICAL PROPERTIES

D. WANG, D. DONETSKY, Y. LIN, G. KIPSHIDZE, L. SHTERENGAS and G. BELENKY[*]

Department of ECE, Stony Brook University, Stony Brook, New York 11794
[*]garik@ece.sunysb.edu

W. L. SARNEY and S. P. SVENSSON

U.S. Army Research Laboratory, 2800 Powder Mill Rd, Adelphi, Maryland 20783

GaInSb and AlGaInSb compositionally graded buffer layers grown on GaSb by MBE were used to develop unrelaxed InAs$_{1-x}$Sb$_x$ epitaxial alloys with strain-free native lattice constants up to 2.1% larger than that of GaSb. The in-plane lattice constant of the strained top buffer layer was grown to be equal to the native, unstrained lattice constant of InAs$_{1-x}$Sb$_x$ with given x. The InAs$_{0.56}$Sb$_{0.44}$ layers demonstrated a photoluminescence (PL) peak at 9.4 μm at T = 150 K. The minority carrier lifetime measured at 77 K for InAs$_{0.8}$Sb$_{0.2}$ was 250 ns.

Keywords: InAsSb; compositionally graded buffer; MBE; infrared, minority carrier lifetime; reciprocal space mapping.

Introduction

GaSb based III-V materials are widely used in the development of mid- and long-wave infrared optoelectronic devices because of the narrow bandgap and the flexibility in forming heterojunctions with various types of band offsets, i.e. type I, type II staggered or type II broken gap. For device applications, heterostructures with a considerable thickness are preferably grown lattice-matched or nearly lattice-matched to the substrate. Therefore, the device design is restricted by the lattice parameters of commercially available III-V substrates. For example, in the case of GaSb based type I lasers, the content of As in InGaAsSb quantum wells must be high enough to satisfy the conditions of pseudomorphic growth, but high As content in quantum wells severely affects the device performance [1]. In principle, the problem can be addressed by the epitaxial growth of lattice-mismatched materials of the desired lattice parameters.

The key issue in mismatched epitaxy is to minimize the dislocations that penetrate through the epi-structures. In this work, we expand the lattice parameter of the GaSb substrate by growing linearly compositionally graded Ga(Al)InSb buffers, following the approach in [2-3]. The graded strain in the buffer layers facilitates the glide of threading dislocations and reduces the densities of dislocations that propagate through the buffer layer into the device [2]. High quality InAs$_{1-x}$Sb$_X$ layers having non-tetragonally distorted, strain-free lattice parameters were grown on top of the buffer layers with thickness up to 1.5 μm.

InAs$_{1-x}$Sb$_x$ alloys are of special interest, because the bowing effect in the band gap E$_g$ is dependent on the Sb composition, which allows the growth of layers having bandgaps narrower than that in InSb [4-15]. In the second part of the paper, we present the optical properties of non-distorted InAs$_{1-x}$Sb$_x$ alloys grown on linearly compositionally graded Ga(Al)InSb buffers. Strong PL was observed for InAs$_{1-x}$Sb$_x$ alloys in a wide temperature range. A relatively long carrier lifetime was obtained in InAs$_{0.8}$Sb$_{0.2}$ alloys from the PL response to modulated optical excitation.

Growth and Structural Characterization

The heterostructures were grown on GaSb substrates by solid-source molecular beam epitaxy using a Veeco GEN-930 system equipped with As and Sb valved cracker sources. Molecular beam fluxes were measured by an ion gauge positioned in the beam path. The substrate temperature was controlled by a pyrometer, which was calibrated using references such as the III to V enriched surface reconstruction transition, oxide desorption and the melting point of InSb. The compositionally graded 2~3.5 μm thick Ga(Al)InSb buffer layers were grown at temperatures ranging from 460 to 520°C. The growth temperature was maintained near 415°C for the InAsSb layers. The Sb incorporation was controlled by the adjustment of the relative pressure of the As and Sb group V elements as measured by the beam-flux-monitor. The growth rate was about 1 μm per hour. InAs$_{1-x}$Sb$_x$ layers with x = 20, 30 and 44% were grown on GaInSb and AlGaInSb buffers.

The defect distribution in linearly compositionally graded GaInSb and AlGaInSb buffers were characterized by cross-sectional TEM images. Figure 1 shows the XTEM images of structures with either laser or absorber layers grown on top of three different linearly graded buffer layerss;including (a) GaInSb with top In content of 16%; (b) GaInSb with top In content of 30%; (c) AlGaInSb with top Al, Ga and In contents of 75, 0 and 25 %, respectively. The images were taken with a (220) bright field two-beam condition to emphasize the dislocations. In all three structures, the misfit dislocation network was confined in the bottom part (~1.5 μm) of the graded buffers; the topmost portion of the buffers as well as the epi-structures grown onto the buffers is free from misfit dislocations. TEM results did not show any noticeable difference in the dislocation morphology of these two buffer layers or in the laser or absorber layer structures grown on top of them, both appear to be equally efficient in accommodating the misfit strain. From the images, we can estimate that the threading dislocation density is below 10^7 cm^{-2} in the InAs$_{1-x}$Sb$_x$ layers.

The surface morphology was characterized by atomic force microscopy (AFM) in tapping mode (AFM Dimension V). Cross-hatched patterns with crossing lines along the [110] crystallographic directions were observed on all structures. However, structures grown on AlGaInSb buffers showed better surface morphology. Figure 2 (a) and (b) show the AFM amplitude images measured over a 50 by 50 μm area for samples

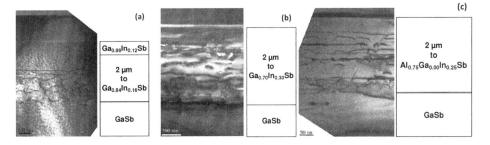

Fig. 1. Cross-sectional TEM images of samples with 2 μm thick linearly graded buffers grown on GaSb substrates: (a) GaInSb with top In content of 16% - mismatch accomodated 0.9%; (b) GaInSb with top In content of 30% - mismatch accomodated 1.4%; (c) AlGaInSb with top Al, Ga and In contents of 75, 0 and 25% - mismatch accomodated 1.4%.

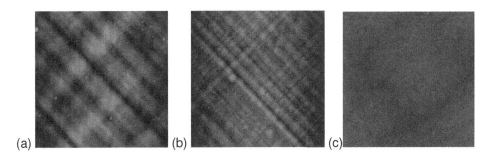

Fig. 2. AFM amplitude images measured over 50 by 50 μm area for samples with InAs$_{0.8}$Sb$_{0.2}$ layer grown on (a) GaInSb buffer and (b) AlGaInSb buffer; (c) shows the enlarged image (3 by 3 μm) of sample (b).

with InAs$_{0.8}$Sb$_{0.2}$ layer grown on a (a) GaInSb buffer and a (b) AlGaInSb buffer. The undulation amplitude and period in sample (a) were about 10 nm and 9 μm, respectively, both nearly twice as much the ~ 5nm and ~ 5 μm measured for sample (b). Figure 2 (c) shows the image of sample (b) measured over 3 by 3 μm area; the root mean square surface roughness, i.e., in between of the dips in cross-hatch pattern, was below 1 nm. Increasing the Sb content led to larger peak-to-peak variations in the cross-hatch pattern, as indicated by surface roughness up to 10 nm for the InAs$_{0.56}$Sb$_{0.44}$ samples.

Strain relaxation of the structures was examined using high-resolution X-ray diffraction reciprocal-space mapping (RSM) at the symmetric (004) and asymmetric (335) Bragg reflections. Figure 3 presents a set of RSM measurements for a structure consisting of a 1 μm InAs$_{0.8}$Sb$_{0.2}$ layer grown on a 2 μm linearly compositionally graded AlGaInSb buffer layer. The native lattice constant of the InAs$_{0.8}$Sb$_{0.2}$ layers is about 0.8% larger than that of GaSb. The native lattice constant of the buffer layer changed from that of GaSb to that of Al$_{0.75}$Ga$_{0.13}$In$_{0.12}$Sb with a strain ramp rate about 0.6% per μm. The topmost section of the graded buffer with Al$_{0.75}$Ga$_{0.13}$In$_{0.12}$Sb composition had a native

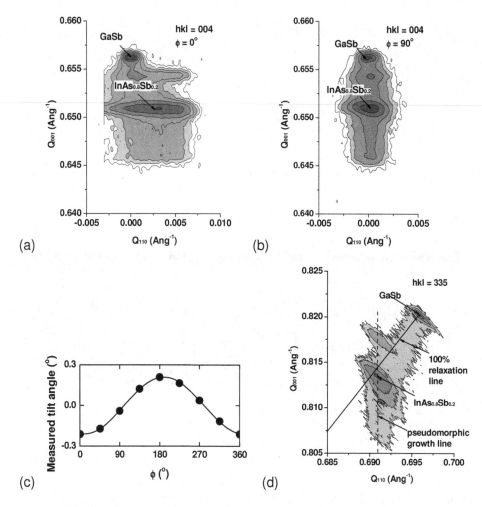

Fig. 3. (a) Symmetric (004) RSM taken at the azimuth angle emphasizing the tilt in the epi-layers; (b) (004) RSM taken at the azimuth angle minimizing the tilt in the epi-layers; (c) dependence of the measured tilt angle as a function of the azimuth angle; (d) asymmetric (335) RSM taken at azimuth angle equal to 90°. Solid line denotes the location of 335 reflexes corresponding to fully relaxed material with lattice parameter gradually increasing from that of GaSb. Dashed line denotes the location of 335 reflexes of the material with further increasing native lattice parameter but grown pseudomorphically to the top of fully relaxed section.

lattice constant about 1.3% larger than that of GaSb, but due to compressive strain, the in-plane lattice constant is equal to the native constant of the bulk $InAs_{0.8}Sb_{0.2}$. When the final structure was grown, the InAsSb layer was sandwiched between $Al_{0.75}Ga_{0.13}In_{0.12}Sb$ carrier confinement layers to assist photoluminescence experiments.

The symmetric reflection revealed the tilt present in the epi-structure. Figure 3 (a) and (b) shows the RSMs obtained near the symmetric (004) reflection at two azimuth angles, namely (a) $\varphi = 0°$ and (b) $\varphi = 90°$, corresponding to two perpendicular [110] crystallographic directions. The tilt angle projected to the measurement plane is determined from the horizontal peak separation between the GaSb substrate and the epi-layers. As shown in Figure 3 (a), the tilt angle increases as the thickness increases in the bottom part of the graded buffer, and stops increasing in the consequent layers. The bottom part of the buffer layers is near completely relaxed, as will be shown later, suggesting that tilting is associated with the process of strain relaxation. Figure 3 (c) plots the projected tilt angle as a function of several azimuth angle φ. We estimate the tilt angle to be 0.2° in the direction about 10° away from the [110] direction ($\varphi = 90°$).

Asymmetric (335) RSM reflexes were measured at four different azimuth angles in order to characterize the degree of relaxation of the graded buffer layer and to confirm that the $InAs_{0.8}Sb_{0.2}$ layer is lattice-matched to the topmost part of the graded buffer. Figure 3 (d) shows one of the (335) RSMs measured at an azimuth angle equal to 90°, i.e., with the minimum tilting effect. The shift visible in the (335) RSM corresponds to the transition from the strain relaxed to the pseudomorphic section of the graded buffer. For illustrative purposes, the solid line corresponds to a 100% relaxed square lattice. The observed relaxation is close to 100%. After the tilt angle is accounted for, the degree of relaxed in this section of the graded buffer can be estimated as 95%, i.e., nearly 100%, and within our experimental error. The pseudomorphic growth of the dislocation-free topmost section of the buffer layer is apparent from the (335) scan since the reflex from the buffer layer is nearly vertical (dashed line in Figure 3 (d)). The reflection from the $InAs_{0.8}Sb_{0.2}$ layer is located at the turning point and on the same vertical line as the pseudomorphic section of the buffer, which confirms lattice matching to the in-plane lattice constant of the graded buffer layer. The amount of strain in the $InAs_{0.8}Sb_{0.2}$ layer is below 0.1%; therefore, no strain relaxation is expected. The reflection located above the InAsSb reflection in both the (004) and (335) RSM corresponds to a pseudomorphically strained auxiliary AlGaSb layer (~150 nm) that was grown on top of the InAsSb layer for calibration purposes.

Optical Characterization

The PL and absorption spectra were measured with a Fourier-transform infrared (FTIR) spectrometer equipped with a liquid-nitrogen cooled HgCdTe detector with a cut-off wavelength of 12 μm. The PL was excited by either a 970 nm laser diode or a Nd:YAG laser and collected by reflective optics. PL was observed from all structures in a wide temperature range, up to room temperature from samples with 20% Sb. Figure 4 (a) shows the PL spectra from 1-μm thick $InAs_{0.8}Sb_{0.2}$ layer grown on an AlGaInSb buffer at 13, 150 and 300 K. Figure 4 (b) presents the PL spectra measured from a 1-μm thick $InAs_{0.56}Sb_{0.44}$ layer grown on an AlGaInSb buffer and a 1.8-μm thick long-wave InAs/GaSb superlattice grown on a GaSb substrate. The superlattice structure consists of 300 periods of InAs and GaSb layers with the cell period of 63 Å enclosed within 20-nm AlAsSb carrier confinement layers lattice-matched to GaSb. The spectral widths (full-width at half maximum) for the three samples were similar, about 11 meV. The PL intensities from both $InAs_{0.56}Sb_{0.44}$ and superlattices were comparable at 13 K while

(a)

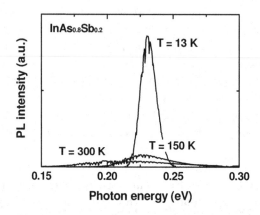

(b)

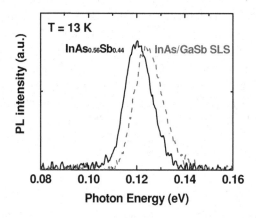

Fig. 4. (a) PL spectra from InAs$_{0.8}$Sb$_{0.2}$ sample grown on AlGaInSb buffer at 13 K, 150 K and 300 K under an excitation power of 0.5W. (b) PL spectra from InAs$_{0.56}$Sb$_{0.44}$ layer grown on AlGaInSb buffer and long-wave InAs/GaSb superlattices grown on GaSb substrate at 13 K under an excitation of 0.1 W. The PL was excitation by a Nd:YAG laser with a beam diameter of about 0.5 mm.

intensities drops much faster in InAsSb sample at elevated temperatures. Considering the challenge of creating adequate hole confinement in the As-rich alloys, the faster drop of the PL intensity with temperature can be explained by the increased diffusion of the excess carriers out of the InAsSb layer.

The absorption spectra were measured for the InAsSb layers with Sb compositions of 20% and 30% grown on GaInSb buffers. The absorbance was determined from transmission measurements taking into account the multiple reflections. The absorption spectrum was derived by subtracting the absorbance of the heterostructure with the epi-layers and the substrate. The transmission of the substrate was determined using the same sample after the epi-layers were removed by polishing. The substrate of the sample with 20% Sb was thinned to 300 μm. The substrate of the heterostructure with 30% Sb was lapped down to near 50 μm thickness because of high free carrier absorption in the GaSb at longer wavelengths. The latter was determined to be 140 cm^{-1} near λ = 8 μm for the GaSb substrates with Te doping level of 3×10^{18} cm^{-3}. The sample with the measured thickness of 55 μm had near 50% transmission at this wavelength, as compared to a 2% transmission for the 300 μm-thick substrate. The free carrier absorption in the thin substrate was determined by a fit based on the absorption measurements for the thicker substrate. Both absorption and PL spectra measured for the two samples at 150 K are presented in Figure 5. The PL peak energy matched the absorption edge indicating that the PL was associated with the band-to-band recombination.

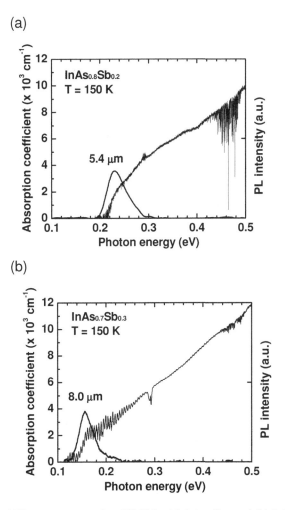

Fig. 5. Absorption and PL spectra measured at 150 K for (a) InAs$_{0.8}$Sb$_{0.2}$ and (b) InAs$_{0.7}$Sb$_{0.3}$. The PL was excited by a 970-nm laser diode at a power of 400 mW; the excitation area was 2.5 x 10^{-3} cm^2. The distortion near 0.29 eV was caused by CO2 absorption.

The energy positions of PL maxima at T = 13 K versus Sb composition x in the InAsSb layers are presented in Figure 6. The positions of PL maxima were used to determine the bandgaps and the bowing parameter, which was about 0.9 eV, considerably greater than the recommended value of 0.7 eV [15]. The lower value of bowing reported previously was based on measurements in materials grown without control of the strain relaxation. The observed difference in the bowing between the 0.9 eV determined in this work and the 0.7 eV reported in literature can be explained by the absence of residual strain in the InAsSb epitaxial layers.

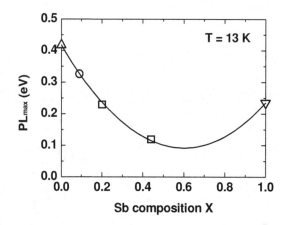

Fig. 6. Dependence of 13K PL maxima on composition X in InAs1-XSbX epitaxial layers: InAs epilayer grown on InAs substrate (triangle), InAsSb0.08 epilayer grown lattice matched to GaSb substrate (circle), InAs1-XSbX epilayers grown on AlGaInSb buffers (squares) on GaSb substrate, InSb epilayers grown on InSb substrate (inverted triangles).

Carrier lifetime measurements for the 1-μm thick InAs$_{0.8}$Sb$_{0.2}$ layer grown on AlGaInSb buffer layer were performed at T = 77 K using optical modulation response technique [16]. To minimize the effects of carrier separation on the carrier lifetime in undoped InAsSb layers, the Al0.25Ga0.70In0.05Sb barriers were doped with Be to the level of 1×10^{17} cm^{-3}. The dependence of the carrier lifetime on the excitation power is shown in Figure 7. Simulation of the band diagram showed that the minority holes remain confined under low excitation (inset of Figure 7 (b)). The carrier lifetime was determined from the PL response to a small signal modulation of excitation in the frequency domain. The PL response spectra in a range of continuous-wave excitation power are shown in Figure 7 (a). The carrier lifetime τ corresponding to the cut-off frequency (-3dB point) was obtained by fitting the response in the entire frequency range to the dependence PL$_\omega \propto [1 + (2\pi f \times \tau^2)^2]^{-1/2}$. A 250 ns carrier lifetime under low excitation condition was measured. The excess carrier concentration was estimated to be in the range $(2\sim4)\times10^{15}$ cm^{-3} at the excitation power in the range of 0.5~1 W/cm^2.

(a)

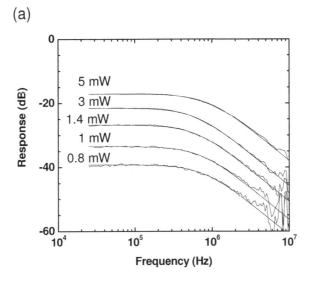

(b)

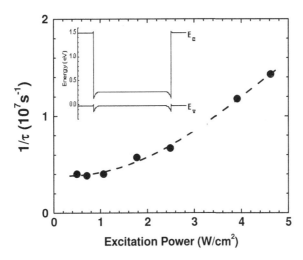

Fig. 7. Carrier lifetime measurements at T= 77 K on a 1-μm-thick InAs$_{0.8}$Sb$_{0.2}$ layer grown on an AlGaInSb buffer on a GaSb substrate. The PL responses and fits are presented for continuous wave excitation power levels of 0.8, 1, 1.4, 3 and 5 mW from bottom to top, respectively. The PL was excited at the wavelength of 1.31 μm, and the excitation area was 2×10^{-3} cm^2 FWHM (left). The reciprocal carrier lifetime is plotted versus continuous–wave excitation power density (right). A schematic band diagram of the InAsSb heterostructure used for carrier lifetime measurements is shown in the inset.

Conclusion

In summary, we conclude that growing compositionally graded buffers (Ga(Al)InSb on GaSb substrates) with a strained but unrelaxed top layer allows the fabrication of bulk $InAs_{1-x}Sb_x$ layers (0.5~1.5 μm thick). These films have characteristics that are promising for the development of IR detectors operating within the spectral range from 5 to 12 μm. The critical element of the technology is the control of the in-plane lattice constant of the topmost section of the buffer. The in-plane lattice constant of this layer must be equal to the lattice constant of $InAs_{1-x}Sb_x$ with given x. The unrelaxed $InAs_{1-x}Sb_x$ epitaxial layers grown on top of such buffers demonstrated photoluminescence in the spectral range from 5.2 to 9.4 μm within the temperature range of 77~150 K. The carrier lifetime of 250 ns was obtained at T = 77 K for structure consisting $InAs_{0.8}Sb_{0.2}$ epi-layers.

Acknowledgement

This work was supported by the National Science Foundation, Grant DMR0710154 and Army Research Office, Grant W911NF11-1-0109. The authors would like to thank IQE, Inc. for providing the superlattice samples.

References

[1] L. Shterengas, G. L. Belenky, J. G. Kim and R. U. Martinelli, *Semicond. Sci. Tech.* **19**, 655 (2004).
[2] J. Tersoff, *Appl. Phys. Lett.* **62**, 693 (1993).
[3] G. Kipshidze, T. Hosoda, W. L. Sarney, L. Shterengas, and G. Belenky, *IEEE Photon. Technol. Lett.* **23**, 317 (2011).
[4] Z. M. Fang, K. Y. Ma, D. H. Jaw, R. M. Cohen, and G. B. Stringfellow, *J. Appl. Phys.* **67**, 7034 (1990).
[5] G. S. Lee, Y. Lo, Y. F. Lin, S. M. Bedair, and W. D. Laidig, *Appl. Phys. Lett.* **47**, 1219 (1985).
[6] Y. B. Li, S. S. Dosanjh, I. T. Ferguson, A. G. Norman, A. G. de Oliveira, R. A. Stradling, and R. Zallen, *Semicond. Sci. Technol.* **7**, 567 (1992).
[7] M. Y. Yen, R. People, and K. W. Wecht, *J. Appl. Phys.* **64**, 952 (1988).
[8] C. G. Bethea, B. F. Levine, M. Y. Yen, and A. Y. Cho, *Appl. Phys. Lett.* **53**, 291 (1988).
[9] J. D. Kim, D. Wu, J. Wojkowski, J. Poitrovski, J. Xu, and M. Razeghi, *Appl. Phys. Lett.* **68**, 99 (1996).
[10] S. R. Kurtz, L. R. Dawson, R. M. Biefeld, D. M. Follstaedt, and B. L. Doyle, *Phys. Rev. B* **46**, 1909 (1992).
[11] T.-Y. Seong, G. R. Booker, A. G. Norman, and I. T. Ferguson, *Appl. Phys. Lett.* **64**, 3593 (1994).
[12] S. Nakamura, P. Jayavel, Y. Kobayashi, K. Arafune, T. Koyama, M. Kumagawa, and Y. Hayakawa, *Semicond. Sci. Technol.* **20**, 1064 (2005).
[13] G. Belenky, D. Denetsky, G. Kipshidze, D. Wang, L. Shterengas, W. L. Sarney and S. P. Svensson, *Appl. Phys.Lett.* **99**, 141116 (2011).
[14] G. Belenky, G. Kipshidze, D. Donetsky, S.P. Svensson, W.L. Sarney, H. Hier, L. Shterengas, D. Wang, and Y. Lin, *Proc. SPIE* 8012, 80120W (2011).
[15] I. Vurgaftman, J. R. Meyer and L. R. Ram-Mohan, *J. Appl. Phys.* **89**, 5815 (2001).
[16] D. Donetsky, S. P. Svensson, L. E. Vorobjev, and G. Belenky, *Appl. Phys. Lett.* **95**, 212104 (2009).

HIGH-PERFORMANCE INTERBAND CASCADE LASERS
FOR λ = 3-4.5 μm

W. W. BEWLEY, C. S. KIM, M. KIM, I. VURGAFTMAN, C. L. CANEDY,
J. R. LINDLE, J. ABELL and J. R. MEYER

Code 5613, Naval Research Laboratory, Washington DC 20375
jerry.meyer@nrl.navy.mil

We discuss the single-mode performance characteristics of midwave-infrared interband cascade lasers. Broad-area devices with 5 active stages display pulsed threshold current densities as low as 375 A/cm^2 and threshold power densities as low as 920 W/cm^2 at room temperature, owing in part to the suppression of Auger recombination. Narrow ridges were processed using optical lithography to incorporate a periodically-corrugated pattern into the sidewalls. The corrugations are intended to suppress lasing in higher-order lateral modes, and also provide a 4th-order grating for distributed feedback. A corrugated-sidewall device operating at T = -20°C produced more than 46 mW of cw output in a single spectral mode, with a peak wallplug efficiency of 7.6%. The device maintains single-mode operation at current densities up to 10 times the lasing threshold, and the single-mode tuning range is 26 nm if both current and temperature are varied. Another device operating at -23°C produces up to 31 mW of single-mode power in the 3.3142-3.3164 μm range that spans several of the strongest absorption signature lines for methane.

Keywords: Interband cascade laser; mid-IR; laser spectroscopy.

1. Introduction

In 2002, a quantum cascade laser (QCL) became the first coherent semiconductor source for the 3-5 μm midwave infrared spectral band to operate in continuous wave mode at room temperature[1]. While strain-balanced InGaAs/InAlAs QCLs emitting at λ ≈ 4.6 μm have recently achieved cw operation to 100°C,[2] wavelengths shorter than 4 μm become increasingly more challenging because excessive strain is needed to assure a sufficient conduction-band offset. The reported limit thus far is λ ≈ 3.8 μm.[3] Conventional type-I quantum-well lasers have also reached room-temperature cw operation in the mid-IR, but only to wavelengths less than ≈ 3.2 μm due to marginal hole confinement[4].

The interband cascade laser (ICL)[5-8] employs the type-II antimonide material system based on InAs electron wells and GaInSb hole wells[9,10]. It exploits a suppression of Auger non-radiative decay[11], and represents a hybrid of the conventional diode and the quantum cascade laser. Even though the lasing transition in its type-II "W" active region is an interband process, multiple stages can be cascaded by energetically aligning the conduction and valence states at the structure's *other* type-II interface, between the electron and hole injectors. This allows electrons in the valence band to scatter elastically back to the conduction band for recycling into the next active stage. At mid-IR wavelengths where current requirements tend to be high, ohmic contributions to the bias voltage can be substantial. This makes the in-series connection of the active wells in a cascade architecture advantageous over the effectively parallel current flow in a conventional multiple-quantum-well laser. The lower current is combined with a higher bias, whose minimum value is the photon energy multiplied by the number of stages.

The first laboratory demonstration of an ICL occurred in 1997[12]. While the early development was limited in part by material growth issues, coupled with an incomplete understanding of the relevant physical processes and design tradeoffs, careful optimization of the design, growth, and processing have led to quite rapid progress since the first NRL ICL growth by molecular beam epitaxy (MBE) in 2005[13-15]. This has led, for example, to our recently surpassing the practical milestone of room-temperature cw operation[16]. Ambient-temperature broad-area ridges have operated in pulsed mode to wavelengths as long as 5.02 μm, with threshold carrier densities in the 390-910 A/cm^2 range for 2.9 μm $\leq \lambda \leq$ 4.5 μm.[17] A recent λ = 3.73 μm device had an even lower threshold current density of 375 A/cm^2 at 27°C, and a corresponding power density threshold of only 920 W/cm^2. A narrow-ridge ICL operated in cw mode to 72°C,[18] which the highest reported for any semiconductor laser emitting in the 3-4 μm range.

Numerous chemical and biochemical sensing applications would benefit from the availability of compact semiconductor sources with narrow spectral output in the mid-IR molecular fingerprint region. While most earlier demonstrations of single-mode ICL emission have been limited to cryogenic temperatures[6,19] or sub-mW powers[20], we recently reported the cw generation of up to 12 mW in a single spectral mode (λ = 3.634 μm) at 25°C.[21] Distributed feedback (DFB) was provided via periodic corrugation of the ridge sidewalls.

The present work discusses a further investigation of this novel approach. The periodic pattern etched into the sidewalls of the narrow ridges plays two roles simultaneously: (1) it should suppress parasitic higher-order lateral lasing modes, which are more concentrated toward the outer ridge boundaries, by preferentially increasing their scattering losses relative to the fundamental mode centered in the middle; and (2) it imposes a 4th-order DFB grating for selection of a single longitudinal mode. One advantage is that the 4th-order grating period is long enough to allow definition by optical lithography, whereas a 1st- or 2nd-order grating requires e-beam lithography. Furthermore, the internal losses normally induced by overlap of the lasing mode with contact metal filling the grooves of a grating etched into the top of the ridge (the GaSb-based system does not provide a convenient overgrowth dielectric) are eliminated by the sidewall grating. It should be noted here that while the in-plane coupling for a symmetric 4th-order grating vanishes at 50% duty cycle, the processed structures contain enough small deviations from ideal symmetry and duty cycle to induce significant interactions between the forward- and backward-propagating waves.

2. Growth and Processing

The five-stage ICL wafers were grown on n-GaSb (100) substrates in a Riber Compact 21T MBE system, using methods discussed elsewhere[22]. The active region designs are similar to those described previously[23], except for a somewhat different GaInSb hole well thickness and composition, and other adjustments to minimize the internal loss. Two bulk GaSb separate-confinement layers (SCLs) were employed above and below the active region. Different wafers were employed for Ridges A and B described below.

The ridges with corrugated sidewalls were fabricated by photolithography and reactive-ion etching (RIE), using a Cl-based inductively-coupled plasma (ICP) process that stopped either within the bottom GaSb SCL (Ridge A) or in the optical cladding layer below that (Ridge B). The ridges were subsequently cleaned with a phosphoric-acid-based wet etch to minimize damage from the ICP RIE. Ridge A, of average width 7.4 μm, had a corrugation amplitude of ≈ 1 μm on each side of the ridge and a period of 2.00 μm to provide a 4th-order DFB grating. Ridge B had an average width of 7.5 μm, and a shorter period of 1.815 μm that was again designed to provide a 4th-order DFB grating for the targeted lasing wavelength. The grating amplitude was again ≈ 1 μm. While Ridge A was processed using a commercial optical lithography mask, the mask used to process Ridge B with custom grating period was fabricated internally by e-beam lithography.

Fig. 1. False-color scanning electron micrograph of a typical processed narrow-ridge ICL with corrugated sidewalls.

A 200-nm-thick Si_3N_4 dielectric layer was deposited onto both structures by plasma-enhanced chemical vapor deposition (PECVD), and the top contact windows were opened by a self-aligned etch back using SF_6-based ICP. Next, 100 nm of SiO_2 was sputtered to block occasional pinholes in the Si_3N_4, followed by e-beam evaporation of the Ag/Ti/Pt/Au contact metallization. The ridges were then electroplated with 5 μm of gold to improve the heat dissipation. The wafers were thinned to ≈ 190 μm, metallized with Ag/Cr/Sn/Pt/Au for the bottom contact, and annealed for 1 minute at 300°C. Cavities were cleaved to lengths of 1.5 mm for Ridge A and 2 mm for Ridge B. A high reflectivity (HR) coating was deposited onto the back facet of Ridge A, by depositing 500 Å Al_2O_3, 1500 Å Au, and 500 Å Al_2O_3. Ridge B had two uncoated facets. The primary difference from the fabrication steps used in Ref. 21 is that the present work introduced the self-alignment process for the top metal contact.

Figure 1 shows a false-color scanning electron micrograph of a typical processed corrugated-sidewall ridge (not one of the two tested for this work). For characterization, the devices were mounted epitaxial-side-up on a copper heat sink attached to the cold finger of an Air Products Heli-Tran Dewar.

3. Lasing Characteristics

Before narrow ridges are patterned, contact lithography and wet chemical etching are used to produce 150-μm-wide lasers for pulsed characterization of each wafer. These broad-area devices are cleaved to a standard cavity length of 2 mm. For operation at 27°C, Wafer A displayed a threshold current density (j_{th}) of 590 A/cm^2, input power density of 1.37 kW/cm^2, differential slope efficiency (*dP/dI*) of 206 mW/A, and peak emission wavelength of 3.70 μm. Wafer B exhibited corresponding characteristics of j_{th} = 482 A/cm^2, 1.19 kW/cm^2, *dP/dI* = 188 mW/A, and λ = 3.45 μm.

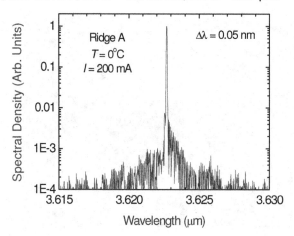

Fig. 2. CW FTIR spectrum of Ridge A at T = 20°C. The linewidth is limited by the spectrometer resolution of 0.05 nm.

High-resolution lasing spectra for Ridge A were obtained with a Bomem DA-8 FTIR. It is apparent from the typical spectrum illustrated in Fig. 2 that when operated at 0°C, the device emitted in a single spectral mode at λ = 3.6238 μm with a sidemode suppression ratio of at least 30 dB. The apparent linewidth of Δλ = 0.05 nm was instrument-limited. Figure 3 shows the corresponding cw light-current (*L-I*) characteristics at a series of temperatures from -20°C to 30°C. While the output was always single-mode at the highest currents shown (up to 10 times the lasing threshold), at the lower temperatures and low currents the spectrum became multi-mode as Fabry-Perot lines began to appear. Presumably this occurred because the gain peak shifted too far away from the DFB grating resonance. The ridge produced 15 mW in a single mode at T = 20°C, with a maximum wallplug efficiency of 3.7%. The power increased to 46 mW at -20°C, where the maximum wallplug efficiency was 7.6%. These single-mode cw powers substantially exceed the requirements of many chemical sensing applications. Note also that for the operating bias voltage of ≈ 2.4 V, the input power required to reach threshold is < 0.1 W at -20 °C and < 0.2 W at 30°C (where a thermoelectric cooler would generally not be required).

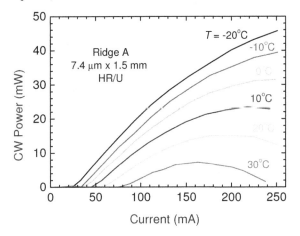

Fig. 3. CW *L-I* characteristics for Ridge A at a series of temperatures between -20°C and 30°C.

Figure 4 illustrates the spectral tuning range of Ridge A as a function of current at a series of temperatures. The output was single-mode under all of the conditions shown. For operation at 0°C the tuning range with current was 11 nm (≈ 0.05 nm/mA, but somewhat nonlinear due to heating and variation of the refractive index), while at the fixed current of 240 mA the tuning range with temperature was 20 nm (40 nm/°C). By varying both temperature and current, one could tune over 26 nm.

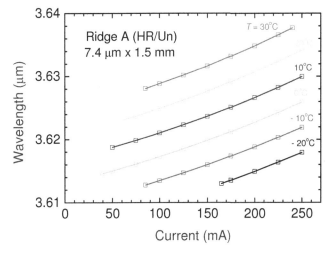

Fig. 4. Wavelength tuning curves for Ridge A at temperatures between -20°C and 30°C and currents up to 250 mA. The output is into a single spectral mode under all the conditions shown.

The ICL wafer employed for Ridge B was selected to emit near some of the strongest methane absorption lines when operated at an intermediate temperature. In this case the emission spectra were acquired with a ½-m grating monochromator. Figure 5 illustrates the spectra at four different currents for a fixed temperature of -27°C. The superimposed red curve (right axis) is a methane transmission spectrum plotted on the same wavelength scale. Tuning of the laser emission from 3.3142 µm to 3.3164 µm is seen to sweep across several of the strongest methane signature lines. The corresponding *I-V* and *L-I* data in Fig. 6 illustrate that a rather high cw power was delivered in this spectral range. Even though this device had no facet coatings, at 400 mA it produced a single-mode output power of 31 mW per uncoated facet. While the nominal input power of 1.2 W (2.6% per facet wallplug efficiency) neglects the power required for cooling, ambient-temperature single-mode powers and efficiencies at least as high as those illustrated above for Ridge A are expected from future devices emitting in this wavelength range.

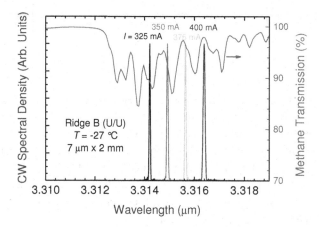

Fig. 5. CW spectra for Ridge B at *T* = -23°C and a series of currents. A transmission curve for methane, showing a series of absorption lines, is plotted on the same wavelength scale.

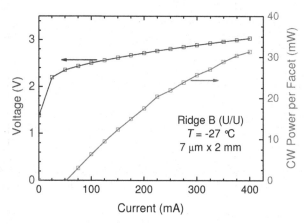

Fig. 6. Single-mode CW *I-V* (left axis) and *L-I* (right axis, per uncoated facet) for Ridge B operating at *T* = -27°C.

4. Conclusions

We have studied interband cascade lasers with ridge sidewalls incorporating a periodic corrugation pattern defined by optical lithography. These devices are shown to be capable of generating high cw output powers in a single spectral mode at temperature in the thermoelectric cooler range. Distributed feedback from the 4th order grating induced by the corrugations provides high-purity spectral emission, with instrument-limited linewidth (≤ 0.05 nm) and > 30 dB sidemode suppression.

The characterization demonstrated single-mode output powers at $\lambda \approx 3.62$ μm of up to 46 mW at -20°C and 15 mW at 20°C. The corresponding wallplug efficiencies were 7.4% and 3.7%, and the required input power was < 0.2 W even at the higher temperature. We anticipate that an optimized thermoelectrically-cooled corrugated-sidewall DFB ICL should require ≤ 0.5 W of total battery power to generate ≥ 10 mW of cw single-mode output. At a slightly lower temperature of -27°C, a second ridge produced up to 31 mW of single-mode cw output in the 3.3142-3.3164 μm region spanning several of the strongest methane absorption lines.

The pulsed characterization data for broad-area lasers processed from existing ICL wafers indicates that, following narrow-ridge processing, ambient-temperature cw operation should be readily attainable at wavelengths spanning 2.9-4.2 μm. With further design optimization it should be possible to extend this level of performance into the 4.5-5.0 μm range. Thus corrugated-sidewall ICLs requiring very low input powers represent a promising new source for mid-IR chemical spectroscopy.

Reference

1. M. Beck, D. Hofstetter, T. Aellen, J. Faist, U. Oesterle, M. Ilegems, E. Gini, and H. Melchior, *Science* **295**, 301 (2002).
2. Y. Bai, S. Slivken, S. R. Darvish, and M. Razeghi, *Appl. Phys. Lett.* **93**, 021103 (2008).
3. M. Razeghi, A. Evans, Y. Bai, J. Nguyen, S. Slivken, S.R. Darvish, and K. Mi, Proc. *Int. Conf. InP and Related Materials* (Matsue Japan, 14–18 May 2007).
4. G. Belenky, L. Shterengas, D. Wang, G. Kipshidze, and L. Vorobjev, *Semicond. Sci. Technol.* **24**, 115113 (2009).
5. R. Q. Yang, *Superlatt. Microstruct.* **17**, 77 (1995).
6. J. L. Bradshaw, J. D. Bruno, J. T. Pham, D. E. Wortman, S. Zhang, and S. R. J. Brueck, *IEE Proc.-Optoelectron.* **150**, 288 (2003).
7. K. Mansour, Y. Qiu, C. J. Hill, A. Soibel, and R. Q. Yang, *Electron. Lett.* **42**, 1034 (2006).
8. C. L. Canedy, C. S. Kim, M. Kim, D. C. Larrabee, J. A. Nolde, W. W. Bewley, I. Vurgaftman, and J. R. Meyer, *J. Cryst. Growth* **301**, 931 (2007).
9. R. H. Miles, D. H. Chow, Y.-H. Zhang, P. D. Brewer, and R. G. Wilson, *Appl. Phys. Lett.* **66**, 1921 (1995).
10. J. R. Meyer, C. A. Hoffman, F. J. Bartoli, and L. R. Ram-Mohan, *Appl. Phys. Lett.* **67**, 757 (1995).
11. W. W. Bewley, J. R. Lindle, C. S. Kim, M. Kim, C. L. Canedy, I. Vurgaftman, and J. R. Meyer, *Appl. Phys. Lett.* **93**, 041118 (2008). The dominant mechanism governing the Auger suppression has not been unambiguously identified.

12. C.-H. Lin, R. Q. Yang, D. Zhang, S. J. Murry, S. S. Pei, A. A. Allerman, and S. R. Kurtz, *Electron. Lett.* **33**, 598 (1997).
13. C. L. Canedy, W. W. Bewley, J. R. Lindle, C. S. Kim, M. Kim, I. Vurgaftman, and J. R. Meyer, *Appl. Phys. Lett.* **88**, 161103 (2006).
14. W. W. Bewley, J. A. Nolde, D. C. Larrabee, C. L. Canedy, C. S. Kim, M. Kim, I. Vurgaftman, and J. R. Meyer, *Appl. Phys. Lett.* **89**, 161106 (2006).
15. W. W. Bewley, C. L. Canedy, M. Kim, C. S. Kim, J. A. Nolde, J. R. Lindle, I. Vurgaftman, and J. R. Meyer, *Electron. Lett.* **43**, 283 (2007).
16. M. Kim, C. L. Canedy, W. W. Bewley, C. S. Kim, J. R. Lindle, J. Abell, I. Vurgaftman, and J. R. Meyer, *Appl. Phys. Lett.* **92**, 191110 (2008).
17. I. Vurgaftman, C. L. Canedy, C. S. Kim, M. Kim, W. W. Bewley, J. R. Lindle, J. Abell, and J. R. Meyer, "Mid-infrared interband cascade lasers operating at ambient temperatures", *New J. Phys.* **11**, 125015 (2009).
18. W. W. Bewley, C. L. Canedy, C. S. Kim, M. Kim, J. R. Lindle, J. Abell, I. Vurgaftman, and J. R. Meyer, *Opt. Engr.* (in press).
19. C. S. Kim, M. Kim, W. W. Bewley, C. L. Canedy, J. R. Lindle, I. Vurgaftman, and J. R. Meyer, *IEEE Photon. Technol. Lett.* **19**, 158 (2007).
20. R. Q. Yang, C. J. Hill, K. Mansour, Y. M. Qiu, A. Soibel, R. E. Muller, and P. M. Echternach, *IEEE J. Select. Top. Quantum Electron.* **13**, 1074 (2007).
21. C. S. Kim, M. Kim, J. R. Lindle, W. W. Bewley, C. L. Canedy, J. Abell, I. Vurgaftman, and J. R. Meyer, *Appl. Phys. Lett.* **95**, 231103 (2009).
22. C. L. Canedy, J. Abell, W. W. Bewley, E. H. Aifer, C. S. Kim, J. A. Nolde, M. Kim, J. G. Tischler, J. R. Lindle, E. M. Jackson, I. Vurgaftman, and J. R. Meyer, submitted to *J. Vac. Sci. Technol. B* **28**, C3G8 (2010).
23. W. W. Bewley, J. R. Lindle, C. S. Kim, M. Kim, C. L. Canedy, I. Vurgaftman, and J. R. Meyer, *Appl. Phys. Lett.* **93**, 041118 (2008).

GaN BASED 3D CORE-SHELL LEDs

XUE WANG, SHUNFENG LI, SÖNKE FÜNDLING, JIANDONG WEI, MILENA ERENBURG,
JOHANNES LEDIG,HERGO H. WEHMANN and ANDREAS WAAG

*Institute of Semiconductor Technology, Braunschweig University of Technology,
38106 Braunschweig, Germany*

WERNER BERGBAUER, MARTIN MANDL, MARTIN STRASSBURG
and ULRICH STEEGMÜLLER

*Osram Opto Semiconductors GmbH, 93055 Regensburg, Germany
xue.wang@tu-bs.de*

As a promising novel route towards highly efficient optoelectronic devices, GaN based 3D core-shell light emitting diodes (LEDs) have attracted increased attention in recent years. In comparison to conventional 2D thin film LED, the 3D LED systems using a core-shell geometry with high aspect ratio are a breakthrough in technology with lots of advantages. In this paper, we review our developed growth strategies of these LED systems on Si and sapphire substrates. A catalyst free selective area growth of GaN 3D core-shell LED systems was realized using patterned substrates by metal organic vapor phase deposition in a convenient continuous-flux growth mode. We have recently suggested that the surface polarity plays a crucial role for the morphology of GaN 3D structure growth. In order to analyze the surface polarity of 3D submicron or micron structures, Kelvin probe force microscopy and selective etching techniques have been developed. During the growth of GaN submicron rods and micron columns on patterned SiO_x/sapphire templates, mixed polarity effects could be detected. A new "truncated pyramid + column" growth method was developed to effectively avoid the formation of mixed polarity and realize single N-polar GaN 3D devices. Transmission electron microscopy and spatially and spectrally resolved cathodoluminescence measurements evidently prove the core-shell structure of 3D LED systems.

Keywords: GaN; LED; 3D; core-shell; MOVPE.

1. Introduction

In recent years, GaN based light emitting diodes (LED) have been considered to be the future for energy-saving and sustainable general lighting applications. In particular, GaN based 3D core-shell LED systems have gained increased attention as a promising novel route towards highly efficient optoelectronic devices [1]. In a 3D core-shell LED structure, it is an interesting aspect that the inner core is n-type GaN, the outer shell builds the p-type part and InGaN/GaN multi quantum wells (MQW) are located between these two types, i.e. the MQWs grow on the whole surface of the inner core (cf. Fig. 1). One of the major advantages of this system is a drastically increased active light emitting area, which can be increased by a factor of about 4* aspect ratio. As a consequence, the total light intensity emitted from the same substrate area [2] is also increased by the same factor, provided that high efficiency 3D LEDs can be fabricated, see equation 1

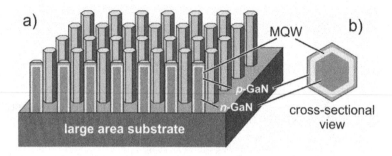

Fig. 1. Core-shell strategy for the fabrication of GaN based 3D LEDs. Sketches of a) core-shell 3D LED ensemble. b) Cross-sectional view of a core-shell 3D LED. The active LED area can be increased approximately by a factor of 4 times the aspect ratio in comparison to a planar LED. Reprinted with permission from S. F. Li, A. Waag, Accepted by Appl. Phys. Rev. Copyright © 2012, American Institute of Physics.

(considering the filling factor F -is taken into account). This could lead to a more effective usage of substrates and processing resources. In addition, the current densities at a constant total current can be reduced, which is one route to circumvent the problem of droop. Moreover, the MQWs cover the whole 3D structures mainly on non-polar (m-plane). In this case, the piezoelectric polarization field is absent [3], which may lead to higher efficiency and wavelength stability.

$$\frac{A_{core}}{A_{film}}\, F = \frac{2\pi r \cdot h}{r^2 \pi}\, F = \frac{2h}{r}\, F$$
$$= 4\,(\text{aspect ratio})\, F \tag{1}$$

The conventional GaN 2D film LEDs suffer from a lack of low cost and large-size native substrates, which causes crystal defects related to lattice and thermal mismatch with the substrates. This substrate problem could be overcome by 3D LED systems, since the GaN nanorods, micron columns or micron pyramids have a small footprint on the substrate which leads to a reduced influence of lattice and thermal mismatch [4]. Therefore, the growth of high quality GaN with 3D structure can be realized on the substrates such as Si or sapphire, especially on large size substrates. A reasonable dimension of a 3D core-shell LED is suggested to be in the order of 500nm diameter or larger and the corresponding aspect ratio larger than 2 [2], to suppress the influence of surface depletion layer. Therefore, submicron or micron sized 3D structures might be more suitable for core-shell LED devices than nanorods.

In recent years, most GaN based LEDs have been grown on sapphire substrates rather than on silicon, because LEDs on Si face the problem of a large mismatch in the thermal expansion coefficients, which leads to a drastic increase of strain during cooling from growth to room temperature. This strain problem has recently been solved by applying strain compensating layers [5, 6, 7]. At the beginning of 2012, OSRAM announced that GaN LED chips on 150mm silicon wafers have been fabricated involving a transfer to a silicon carrier and a removal of the original silicon substrate [8]. This means in the coming years Si substrate technology will become more competitive for GaN based LEDs.

Ga-polar GaN columns and columnar LEDs have been successfully realized in the past by molecular beam epitaxy (MBE) [9, 10, 11] as well as by metal organic vapor phase epitaxy (MOVPE) on Ga-polar GaN templates [12, 13]. Recently, many groups have suggested that N-polar GaN presents potential benefits due to the reversed direction of polarization, which can be used in many novel device structures such as high carrier injection LED [14], high electron mobility transistors [15] and photovoltaic devices [16]. In this paper, we review our understanding of the surface polarity influence on GaN 3D growth and our developed growth strategies of these core-shell LED systems on pre-patterned Si and sapphire substrates. We have demonstrated that polarity plays a very critical role for the morphology of GaN 3D structures, and under hydrogen carrier gas the vertical growth with high aspect ratio of N-polar GaN column is "easier" than Ga-polar structure [17]. On nitrided patterned SiO_x/sapphire or SiN_x/sapphire templates, 3D GaN columns with a mixed polarity (both N- and Ga-polarity) were often observed when employing a continuous growth mode [17, 18, 19]. Mixed polarity structures include inversion domain boundaries (IDB) between the Ga- and N-polar parts, which are expected to cause high reverse leakage currents [20] and hence to reduce the efficiency of LEDs containing IDBs. Therefore, single polar GaN columns are expected to be a prerequisite for high efficiency 3D core-shell LEDs. We have developed a new "truncated pyramid + column" growth method which can effectively avoid the formation of mixed polarity and realize selective area growth of single N-polar GaN columns.

2. GaN Based 3D LEDs on Si Substrate

The position and size controlled growth of GaN submicron or micron 3D structures on Si substrate has become a very attractive topic in recent years, especially for MOVPE growth, due to low cost and high thermal conductivity of Si substrate. Furthermore, the original Si substrates might be removed after GaN LED growth, thus the light absorption from Si could be reduced. In this paragraph, the first part deals with GaN 3D LED structures on etched Si pillar arrays, whereas in the second part we describe the selective area growth (SAG) of GaN 3D structures on several pre-patterned Si templates.

2.1. *GaN based 3D core-shell LEDs on deep etched Si substrate*

We have realized the growth of GaN 3D core-shell LED on etched Si pillar arrays. Figures 2 (a) and (b) show a scanning electron microscope (SEM) image and a sketch of these structures. The template with deep etched Si pillar arrays has been produced by e-beam lithography (Vistec EBPG 5000 + system, 50kV) for patterning of the Si template with a subsequent inductively coupled plasma (ICP) etching (Sentech instruments 500C). The depth of the Si pillars is 2-8μm and the diameter varies from 500nm to 5μm. For the subsequent growth of the LED structure we employed a vertical 3 × 2" FT Thomas Swan MOVPE system. Before the growth, the patterned substrates were thermally cleaned at

1060°C in a hydrogen atmosphere. The 3D LED growth started with an 100nm AlN nucleation layer which is followed by n-doped GaN column growth, a core-shell fivefold InGaN/GaN MQWs and 200nm p-doped GaN cover layer (more experimental details can be found in ref. 21). It should be emphasized that the AlN und GaN growth also occurs on both the side walls of the pillars and the bottom between the pillars, since the bottom was not passivated by a respective mask material. The reason for the occurrence of a pyramidal shape with r-planes instead of a column with m-planes will be explained later in more detail. Our experimental observations show that the growth rate on top of the pillar is higher and at the bottom lower, if the pitch is smaller. This result might be explained by the following model: the growth rate on the top surface of an isolated and small pillar serves as a kind of sensor for the concentration of reactive species in the gas phase. When this top surface extends into the region where the concentration of reactive species in the gas phase is reduced due to the diffusion towards the bottom growth front, the growth rate on the top surface will be reduced. This model can be used in an ensemble of pillars with decreasing pitch, the concentration of reactive species increases, because the influence of the bottom surface material sink is decreasing.

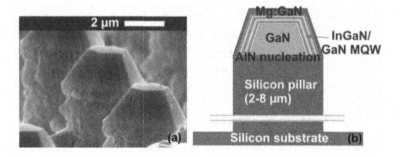

Fig. 2. (a) Side view SEM image of a GaN LED structure on a 5µm Si pillar array; (b) sketch of cross-sectional setup of the GaN LED structure as shown in (a). Reprinted with permission from S. Fündling, Ü. Sökmen, A. Behrends, M. A. M. Al-Suleiman, S. Merzsch, S. F. Li, A. Bakin, H.-H. Wehmann, A. Waag, J. Lähnemann, U. Jahn, A. Trampert, H. Ricchert, phys. stat. sol. (b), 247 (2010) 2315 2328. Copyright © 2010 WILEY-VCH Verlag GmbH & Co. KGaA, Weinheim.

Spatially resolved cathodoluminescence (CL) was employed to characterize the grown GaN 3D core-shell LED structures. Figure 3 presents CL spectra of a single GaN pyramid excited by 10kV electron beam on top and side facet. The measurements were performed at 6K and show GaN near band edge luminescence (360nm), donor-acceptor pair transition (382nm) with its LO phonon replicas, the MQW emission on sidewall (407nm), on top facet (500nm) and yellow luminenscence (560nm) from defects.

On different facets, i.e. top surface and side wall, the thickness and in concentration of the embedded MQWs might be different. This may be due to the different diffusion coefficients of In and Ga species. Besides, polarity effects can play an important role on the different crystal planes. Therefore the CL signals on different facets are also different.

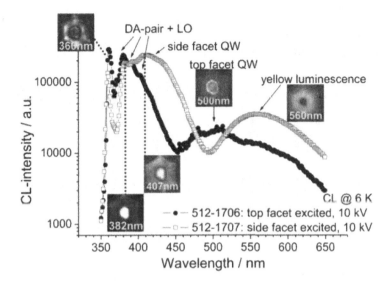

Fig. 3. CL spectra of a single GaN pyramid excited by electrons with an acceleration voltage of 10 kV on top and side facet, respectively. The insets show the corresponding CL images of a 5mm GaN LED structure recorded at the indicated wavelengths. Reprinted with permission from S. Fündling, Ü. Sökmen, A. Behrends, M. A. M. Al-Suleiman, S. Merzsch, S. F. Li, A. Bakin, H.-H. Wehmann, A. Waag, J. Lähnemann, U. Jahn, A. Trampert, H. Riechert, phys. Stat. Sol. (b), 247 (2010) 2315-2328. Copyright © 2010 WILEY-VCH Verlag GmbH & Co. KGaA, Weinheim.

2.2. *Growth of GaN 3D structure on Si substrate*

Besides the deep etched Si pillar template, we have also explored the growth of 3D GaN structures on various Si templates. Figures 4 (a)-(d) show four different templates which we have employed to successfully grow single crystal 3D GaN structure and related core-shell LEDs. In Figures 4 (a) and (b), the mask pattern with an array of hexagonal openings was at first transferred onto the SiO_x on Si or SiO_x on AlN layer on Si substrate by photolithography, and then these openings were etched in the SiO_x layer using an inductively coupled plasma (ICP) etcher. The subsequent growth of the GaN 3D structures started from a nucleation in the opening on Si or AlN respectively. Figure 4 (c) presents a GaN column growth on nitrided Al islands. To produce this template, the mask pattern with an array of hexagonal openings was transferred directly onto Si by photolithography, subsequently 5nm Al was evaporated onto the pattern. After lift-off of the photo resist, the Si template with Al islands was in-situ thermally baked and nitrided. During the nitridation process Al islands were transformed to AlN and thin SiN_x formed between these islands, which served as passivation layer for the subsequent GaN growth. Thus, selective area growth of GaN columns on this template is possible. However, the homogeneity of GaN columns is still not ideal and further optimization processes has to be done. The GaN columns growth on deep etched AlN islands is shown in Figure 4 (d). Before growth, a nitridation on the templates was employed to passivate Si between islands with SiN_x layer. The crystal quality, homogeneity and selectivity of deep etched

AlN islands is much better than the nitrided AlN shown in Figure 4 (c). The morphology of GaN 3D structure on this template is pitch dependent. With increasing pitch, the vertical growth is more visible.

The polarities of these 3D GaN structures shown in Figure 4 (a)-(d) are all Ga-polar, which was proven by KOH wet etching and Kelvin probe force microscopy (experimental details can be found in paragraph 3). We assume that this is the reason why a pronounced and controllable vertical growth is difficult to realize in this case.

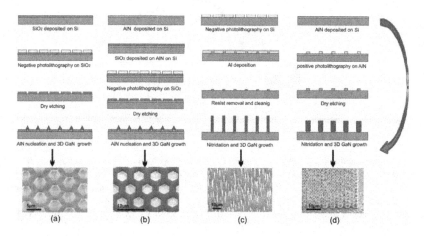

Fig. 4. Main processes involved in the selective area growth: (a) GaN pyramids grow on negative patterned SiOx/sapphire template; (b) GaN pyramids grow on negative patterned SiOx/AlN/sapphire template; (c) GaN columns grow on nitrided 5nm thick Al islands, the template is produced by photoresist lift-off technique; (d) GaN columns grow on positve etched AlN islands.

3. Growth of GaN 3D Structure on Sapphire Substrate

Sapphire as a favourite substrate has been used in many commercial LEDs, especially in GaN/InGaN LEDs. For GaN 3D LEDs, various groups reported on hole-patterned SiO_x (or SiN_x)/GaN/sapphire templates in a submicron or micrometer size. However, in most cases pyramidal structures are observed on these templates by MOCVD [22, 23, 24]. In order to understand how to achieve vertical growth of GaN 3D structures with larger active surface in a standard continuous-flux mode, different carrier gases have been employed and selective area growth of GaN submicron or micrometer structures on patterned SiO_x/sapphire, SiO_x/GaN/sapphire, and SiO_x/bulk GaN templates have been performed. Mixed polarity effects were often observed when using nitrided patterned SiO_x/sapphire or SiN_x/sapphire templates. With the purpose of clarifying its origin, the spatial distribution of surface polarity was analyzed by both Kelvin probe force microscopy and selective etching techniques. The information on polarity has lead to a new "truncated pyramid + column" growth method to effectively avoid the formation of mixed polarity and realize single N-polar GaN columns on patterned SiO_x/sapphire template.

3.1. *Carrier gas, polarity and its influence on growth of GaN 3D structure*

Recently, we demonstrated a more conventional continuous-flux growth of GaN submicron rods directly on patterned sapphire with SiO_x mask layer. H_2 concentration in the carrier gas was found to be a critical parameter in shaping the morphology of GaN submicron rods [19, 25]. If pure N_2 as carrier gas was employed, only pyramidal structures were observed (Fig. 5 (a)). With increasing H_2 fraction in the carrier gas, the vertical growth of GaN submicron rods is stronger and the height of GaN submicron rods increases (Fig. 5 (b)-(c)). Instead, the filling of the openings becomes more inhomogeneous, i.e. some holes were not filled with GaN. Furthermore, the diameter of the GaN submicron rods was found to decrease with increasing H_2/N_2 ratio.

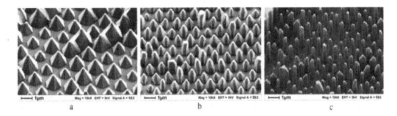

Fig. 5. (a) Growth with pure nitrogen as carrier gas leads to pyramidal-shaped GaN growth, (b) whereas submicron rods growth occurs with H_2/N_2 carrier gas mixture of 1/2 and (c) still improves for H_2/N_2 carrier gas mixture of 2/1. Reprinted with permission from W. Bergbauer, M. Strassburg, Ch. Kölper, N. Linder, C. Roder, J. Lähnemann, A. Trampert, S. Fündling, S. F. Li, H.-H. Wehmann and A. Waag Nanotechnology 21 (2010) 30520. Copyright © 2010 Institute of Physics.

The origin of the influence of H_2 in shaping the morphology of GaN submicron rods or micro columns has been clarified during the investigation of the influence of the polarity on GaN 3D structure growth. Figure 6 presents the morphology of GaN microstructure grown on patterned SiO_x /N-polar bulk GaN, SiO_x/GaN/sapphire (Ga-polar) and SiO_x /sapphire templates. For SiO_x /sapphire templates, a thermal bake and a nitridation step under NH_3 atmosphere were performed before the GaN growth to enable N-polar GaN growth in the openings [26]. For patterned SiO_x /GaN/sapphire and SiO_x /N-polar bulk GaN templates, the temperature was directly ramped up to growth temperature of 1080°C with NH_3 and N_2 carrier gas. Afterwards, trimethylgallium (TMGa) was introduced into the reactor for GaN 3D structure growth. The V/III ratio was kept at about 100 and the flow ratio of H_2/N_2 carrier gas was kept at 2:1 for all samples. In addition, silane was injected into the reactor during the column growth for n-type doping [27]. The growth patterns were defined by nanoimprint or photo- lithography and inductively coupled plasma dry etching.

Figures 6 (a) and (b) show distinctly different morphologies of GaN 3D structures grown on patterned N-polar and Ga-polar GaN substrates. On N-polar GaN template, a clear columnar structure with vertical sidewalls can be observed (cf. Fig. 6 (a)), which is substantially different from pyramidal GaN structures grown on Ga-polar GaN

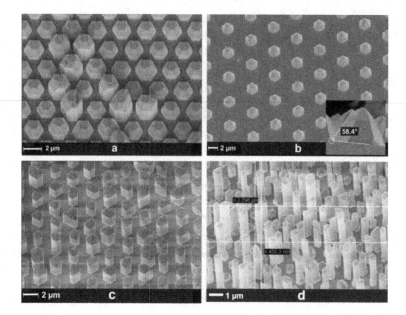

Fig. 6. Comparison of GaN sub-μm rods grown on different templates with the same growth parameters. The images were obtained by tilting the samples 30° with respect to the sample normal. a) patterned SiO$_x$ covered N-polar bulk GaN template; b) patterned SiO$_x$/Ga-polar GaN/sapphire template, GaN 3D structures typically show a pyramidal shape. inset: detailed side view of one structure; c) patterned SiO$_x$/sapphire templates; d) patterned SiO$_x$/sapphire templates with smaller opening size of 400 nm, the measured rod shows a diameter of about 460 nm and a height of about 5.6 μm, yielding in aspect ratios up to 12. Reprinted with permission from S. F. Li, S. Fuendling, X. Wang, S. Merzsch, M. A. M. Al- Suleiman, J. D. Wei, H.-H. Wehmann, A. Waag, W. Bergbauer, and M. Strassburg, Crystal Growth & Design, 11 (2011) 1573. Copyright © 2011 American Chemical Society.

substrate (cf. Fig. 6 (b)). The inset shows a cross-section view of a typical pyramidal structure. The angle between bottom (0001) surface and edge line between two adjacent planes is 58.4°, indicating that these planes are {10-11} planes, which fits the results shown in Figure 2. Those stable planes limit the final morphology of the GaN crystals [28]. On nitrided SiO$_x$/sapphire templates, GaN submicron rods and micron columns can also be realized using the same growth parameters as on Ga-polar template (Fig. 6 (c) and (d)). Both KPFM and wet chemical etching results proved the polarities in these 3D GaN structures [29].

In general, N-polarity determines the vertical growth in continuous-flux, finally leading to GaN submicron rods and micron columns. Besides, the growth on unstrained N-polar and Ga-polar GaN substrates indicate, that strain does not play a critical role for the development of GaN 3D structures. In order to explain the influence of both polarity and H$_2$ carrier gas concentration on the morphology of GaN 3D growth, we have proposed the following growth model. It has been reported in several publications that hydrogen can passivate the N-terminated surfaces, i.e. {000-1} (N-polar surface) and {10-11} r-plane surfaces [30, 31, 32]. The atomic structure of GaN with Ga-polar and N-polar surfaces are depicted in Figures 7 (a) and (b). If H$_2$ is used in MOVPE growth as carrier gas, there are plenty of hydrogen atoms in the growth ambience. Thus, {10-11} r-planes during Ga-polar GaN 3D structure growth could be passivated with N-H bonds [31] hence a low growth rate on these planes (indicated in Fig. 6 (a)). According to Wulff's theory [33] the crystal facets with smaller growth rates mainly determined the shape of the structures, and in this case lead to pyramids. In contrast, the {10-1-1} r-plane surfaces in an N-polar GaN 3D structures are terminated by Ga atoms (Fig. 6 (b)), so passivation is not expected to occur. The larger growth rate reduces the contribution of r-planes to a point where no r-planes are visible any more (Fig. 7 (c)).

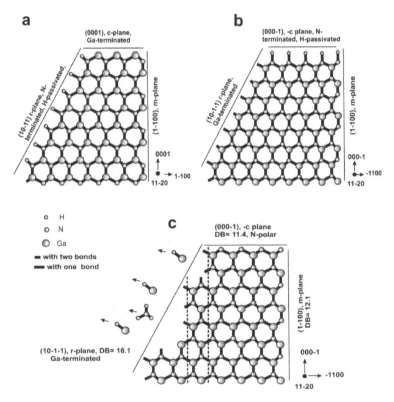

Fig. 7. Schematic drawings of GaN atomic structure. a) GaN pyramidal structure with Ga-polar top surface. b) GaN pyramidal structure with N-polar top surface. c) N-polar GaN pyramidal structure etched by hydrogen. Dashed lines indicate the more stable (1-100) M- plane. The detailed surface reconstructions are not included in this schematic drawing. Reprinted with permission from S. F. Li, S. Fuendling, X. Wang, S. Merzsch, M. A. M. Al- Suleiman, J. D. Wei, H.-H. Wehmann, A. Waag, W. Bergbauer, and M. Strassburg, Crystal Growth & Design, 11 (2011) 1573. Copyright © 2011, American Chemical Society.

These experiments demonstrated that with hydrogen carrier gas the vertical growth of N-polar GaN 3D structures is more straightforward than on Ga-polar GaN. Nevertheless, it is possible to achieve Ga-polar GaN columns by MOVPE growth if one can modify the r-plane surface stability under hydrogen exposure, e.g. by exposing a Ga flux to the 3(N-H) passivated surface which can lead to an unstabilization of this surface [31].

Hersee et al. reported for the first time a successful growth of Ga-polar GaN nanorods in a pulsed growth mode [12]. During this 3D growth process, the Ga-precursor and NH_3-flow were alternatively pulsed into the reactor after nucleation step, which might disturb the 3(N-H) passivation on Ga-polar r-plane and thus leads to a pronounced vertical growth.

3.2. *Mixed polar GaN columns and Polarity analysis by photo-assisted Kelvin probe force microscopy*

3D GaN columns with a mixed polarity (both N- and Ga-polarity) were often observed when employing continuous growth mode by MOVPE on nitrided patterned SiO_x/sapphire or SiN_x/sapphire templates [17, 18, 19]. In order to clarify the origin of the occurrence of mixed polarity, surface photovoltage measurements have been performed by KPFM in addition to etching experiments, analyzing the polarity of GaN during the nucleation stage.

The 3D GaN columns on patterned SiO_x/sapphire or SiN_x/sapphire templates were grown in two steps. The first step was a nucleaton step that was performed at 1000°C and a H_2/N_2 flow rate ratio of 1:1, to ensure a high selectivity of nucleation. In the second GaN column growth step, in order to enhance the vertical growth of GaN columns, the growth temperature and the flow rate ratio were increased to 1080°C and 2:1 respectively. In this step, silane was injected into the reactor for n-type doping. A SEM image of a nucleation seed after 100 s growth is shown in Fig. 8 (a). The dashed hexagon marks the contour of the aperture. N-polar seed lies inside the dashed hexagon and Ga-polar material regularly occurs outside of the aperture. This indicates that the mixed polarity emerges in the nucleation step. The occurrence of Ga-polar GaN is obviously due to additional nucleation seeds on the SiO_x mask. The nucleation step was performed at a low flow ratio of H_2/N_2 carrier gas and a low growth pressure that leads to a low vertical and high lateral growth rate of both polarities [19, 34, 35]. Under the described growth conditions, we have observed that the top surface c-plane of both polarities grows slower than the other facets. According to the Wulff's theory [33], the c-facets of Ga-polar and N-polar domains appear during the nucleation step.

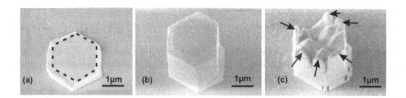

Fig. 8. 30°-tilted SEM images of (a) GaN nucleation seed, where the dashed hexagon indicates the contour of the aperture in the SiO_x layer; (b) GaN column grown on the nucleation seed shown in (a); (c) the GaN column shown in (b) after 5 min. of etching in hot KOH (the arrows indicate Ga-polar domains). Reprinted with permission from Xue Wang, Shunfeng Li, Sonke Fundling, Jiandong Wei, Milena Erenburg, Hergo-H. Wehmann, Andreas Waag, Werner Bergbauer, Martin Strassburg, Uwe Jahn and Henning Riechert, Crystal Growth & Design, 12 (2012) 2552. Copyright © 2011, American Chemical Society.

Vertical growth of GaN column as the second step started on the c-facets of both the Ga- and N-polar material [cf. Fig. 8 (b)]. The orientations of nucleation seeds are obviously preserved in the polarity domains of GaN columns. Figure 8 (c) presents the etching result: N-polar domains display hillock-like surfaces, while Ga-polar domains remain intact. The polarity distribution of a GaN column which was etched by 2M KOH at 80°C for 10 min [36] (not etched down to sapphire) was analyzed by photo-assisted KPFM. The evolution of surface photovoltage (SPV) of Ga- and N-polar domains is shown in Fig. 9 and the topographic image of the etched GaN column is shown in the inset. After switching on 360nm UV light, the initial SPV signals for both polarities rise immediately. For the duration of the illumination, the SPV of the intact domain decreases very slowly, while the SPV signal of the etched down GaN surface decreases rapidly in a logarithmic way. In comparison the KPFM results on Ga- and N-polar GaN layer [37], the SPV behaviors on different polarity domains in a GaN column is in agreement with the KOH etching result.

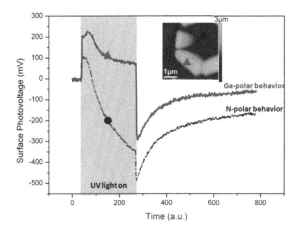

Fig. 9. Evolution of surface photovoltage of the surface of Ga- and N-polar domains in one KOH etched GaN column under 360nm UV light illumination at UV light intensity of 7μW/cm². Inset: Topographic image of the etched GaN column. Reprinted with permission from Xue Wang, Shunfeng Li, Sonke Fundling, Jiandong Wei, Milena Erenburg, Hergo-H. Wehmann, Andreas Waag, Werner Bergbauer, Martin Strassburg, Uwe Jahn and Henning Riechert, Crystal Growth & Design, 12 (2012) 2552. Copyright © 2011, American Chemical Society.

3.3. *Growth of single nitride polar GaN columns*

A new "truncated pyramid + column" growth method has been developed to suppress the formation of mixed polarity in GaN columns on pre-patterned SiO_x/sapphire or SiN_x/sapphire templates. This two-step approach is schematically shown in Fig. 10. The growth of truncated pyramids as the first step is performed at a reduced growth temperature of 960°C instead of 1000°C [cf. Fig. 10 (a)]. It has been observed that the growth rate of the r-planes of Ga-polar GaN decreases and the in <0001> and <1-100> directions increases with decreasing temperature [34, 38]. During this nucleation step at a

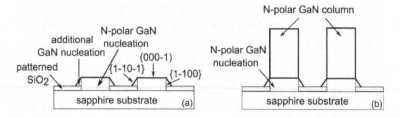

Fig. 10. Schematic representation of the "truncated pyramid + column" growth approach: (a) Growth of GaN truncated pyramids with single N-polar top surface; (b) Growth of single N-polar GaN columns. Reprinted with permission from Xue Wang, Shunfeng Li, Sonke Fundling, Jiandong Wei, Milena Erenburg, Hergo-H. Wehmann, Andreas Waag, Werner Bergbauer, Martin Strassburg, Uwe Jahn and Henning Riechert, Crystal Growth & Design, 12 (2012) 2552. Copyright © 2011, American Chemical Society.

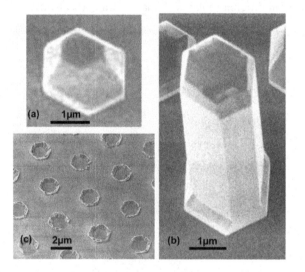

Fig. 11. 30°-tilted SEM images of (a) GaN truncated pyramid after first nucleation growth step, (b) single N-polar GaN column, (c) array of single N-polar GaN columns after 15 min. etching in hot KOH. Reprinted with permission from Xue Wang, Shunfeng Li, Sonke Fundling, Jiandong Wei, Milena Erenburg, Hergo-H. Wehmann, Andreas Waag, Werner Bergbauer, Martin Strassburg, Uwe Jahn and Henning Riechert, Crystal Growth & Design, 12 (2012) 2552. Copyright © 2011, American Chemical Society.

low growth temperature of 960°C, the growth rate of the r-planes of Ga-polar seeds is smaller than that of the other two facets. Based on Wulff's theory [33], crystal facets with smaller growth rates mainly determine the shape of a 3D crystal. Thus, during the nucleation step, the {10-11} r-planes of the Ga-polar nucleation seeds around the apertures appear, originating from the Ga-polar nucleation sites on the SiO_x mask. After a certain growth time, Ga-polar top facet (0001) and the m-facets {1-100} of the Ga-polar nucleation seeds merge and at the same time the truncated pyramids are formed with pure N-polar top surfaces. Inside the opening the seeds are N-polar with a flat surface. Figure 10 (b) presents the second step of the column growth on top of the truncated pyramid. The {1-101} r-plane of the Ga-polar nucleation seed can be passivated with N-H bonds [31], hence, the growth rate on this plane is slow [17]. Under these growth conditions, the GaN columns grow exclusively on the N-polar c-facets of the seeds and not on the r-plane of the Ga-polar additional nucleation seeds, which leads to GaN columns of single N-polarity.

Figures 11 (a) and (b) show SEM images of a truncated pyramid nucleation seed with a single N-polar top surface and a single N-polar column, respectively. Due to the lateral growth of the m-planes, the diameter of the column seems larger than the diameter of the truncated pyramid c-plane surface. The SEM image of an ensemble of single N-polar GaN columns after etching for 15 min. in hot KOH is shown in Figure 11 (c). The GaN columns were completely etched away and just the Ga-polar nucleation collars around the apertures still remain on the SiO_x mask. This etching result indicates that the GaN columns are single N-polar, which was also confirmed by KPFM [29].

4. Growth and Characterization of GaN Based 3D Core-Shell LED on Sapphire Substrate

InGaN/GaN MQWs were grown onto the n-doped GaN column with the same growth parameters as those in the 2D layer MQWs growth. A TEM cross-sectional view of a GaN core-shell LED structure is presented in Figure 12. Threefold InGaN/GaN MQWs and p-GaN cover the whole surface of the GaN column, indicating a core-shell type of growth. The thickness of InGaN QW and GaN quantum barriers (QB) on side wall m-plane is about 2 nm and 8 nm respectively.

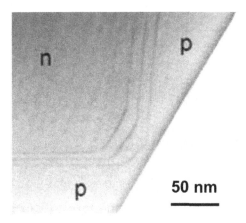

Fig. 12. TEM of a core-shell 3D LED structure grown by MOPVE, including a triple InGaN-GaN quantum well. Core: n-type, Shell: p-type. Growth direction: downward. Reprinted with permission from A. Waag, X. Wang, S. Fuendling, J. Ledig, M. Erenburg, R. Neumann, M. A. M. Al- Suleiman, S. Merzsch, J. D. Wei, S. F. Li, H.-H. Wehmann, W. Bergbauer, M. Strassburg, A. Trampert, U. Jahn, and H. Riechert, Phys. Status Solidi C, 8 (2011) 2296. Copyright © 2011 WILEY-VCH Verlag GmbH & Co. KGaA, Weinheim.

Core-shell LED structures could be realized on mixed polar as well as on single N-polar GaN columns under the same MQWs and p-GaN growth conditions. Room temperature spatially-resolved CL was measured on these core-shell LED structures. Monochromatic intensity maps were taken at wavelengths of 400 nm, 460nm and 480nm respectively using an acceleration voltage of 8 kV. Figures 13 (a-b) and (c-d) show superimposition of such CL maps on corresponding SEM images of a single N-polar and a mixed polar GaN column core-shell LEDs at different wavelengths, respectively. The MQWs emission from m-plane side walls of both single N-polar and mixed polar LEDs was observed with a maximum at 400 nm. However, the dominant MQWs emission from c-plane of single N-polar LED is at about 480nm and of mixed polar LED is at about 460nm.

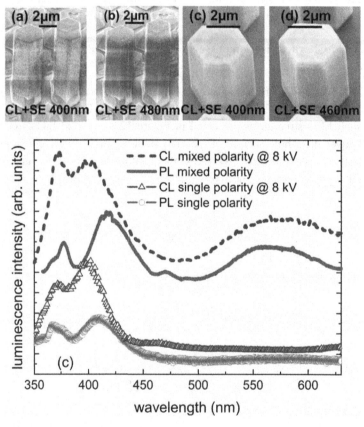

Fig. 13. Colour-coded RT CL maps superimposed on SEM images of single N-polar core-shell LEDs, recorded for the side wall MQW emission at 400 nm (a) and the top facet MQW emission at 480 nm (b) respectively; and mixed polar core-shell LEDs, recorded for the side wall InGaN/GaN MQW emission at 400 nm (c) and top facet MQW emission at 460 nm (d) respectively. (e) Normalized CL and PL spectra at room temperature for single N-polar core-shell LEDs and mixed polar core-shell LEDs; these spectra are vertically shifted for clarity. Reprinted with permission from Xue Wang, Shunfeng Li, Sonke Fundling, Jiandong Wei, Milena Erenburg, Hergo-H. Wehmann, Andreas Waag, Werner Bergbauer, Martin Strassburg, Uwe Jahn and Henning Riechert, Crystal Growth & Design, 12 (2012) 2552. Copyright © 2011, American Chemical Society.

Room temperature PL and CL spectra of single N-polar and mixed polar GaN core-shell LEDs are shown in Figure 13 (e). A pronounced yellow band emission (YL) can be observed in the mixed polar core-shell LEDs, while it is practically absent in the single N-polar core-shell quantum well structures. In contrast to the single N-polar core-shell LEDs, mixed polar LEDs still contain Ga-polar domains and IDBs besides N-polar domains. In Ga-polar domains, Ga vacancies (V_{Ga}) and/or related complexes are widely believed to be YL sources [39]. In addition, the IDBs and corresponding dislocations in the mixed polar LEDs can generate YL [40]. On the other hand, N-polar GaN layers show much lower YL than Ga-polar layers, has been observed by Q. Sun et al. and D. Du et al. [41, 42], which could be explained by a reduction of the formation of V_{Ga} in N-polar GaN as a result of a larger migration length of Ga atoms on a N-polar surface [43, 44]. Therefore, the single N-polar core-shell LEDs show weak YL.

5. Summary and Outlook

In summary, we reviewed our catalyst-free GaN 3D core-shell LED growth on various pre-patterned Si and sapphire templates by MOVPE using a conventional continuous-flux growth mode. Growth mechanisms of GaN 3D structures were extensively discussed. We suggested that the surface polarity plays a crucial role for the morphology of GaN 3D structures, and N-polar GaN columns with high aspect ratio can be achieved in a more straightforward way in comparison to Ga-polar structures. Both N- and Ga-polar GaN domains in one GaN column could be detected when using patterned SiO_x or SiN_x/sapphire templates. The spatial distribution of surface polarity of 3D GaN structures was analyzed by both Kelvin probe force microscopy and selective etching techniques. A new "truncated pyramid + column" growth method was proposed to achieve single N-polar GaN column growth on the same templates used for mixed polar GaN column growth. TEM and CL measurements evidently indicate a core-shell structure of GaN 3D LEDs.

More efforts are required to realize and understand the growth of GaN 3D core-shell LEDs with high aspect ratio on conductive substrates, e.g. Si and GaN templates, which have a better compatibility for high efficiency LED devices. In contract to a traditional 2D layer LED growth, the growth conditions for 3D LED structures in a continuous-flux mode are very different, which have been described in the preceding parts, e.g. V/III ratio and carrier gas. In this case, the defect formation and doping properties in submicron or micron size GaN 3D structures have to be carefully investigated. Furthermore, the processing on 3D LED devices is still an issue which needs further design and optimization for a highly efficient and reliable GaN 3D structure device.

Acknowledgements

This work has been funded partly by the German Ministry of Education and Research (BMBF) within the Project "Monalisa" (Project No. 01BL0811) and partly by the European Community within the FP 7 project "SMASH" (Project No. 228999). We are grateful to D. Rümmler, A. Schmidt and B. Matheis for the substrate preparation and

M. Schilling and F. Ludwig for giving us the possibility to use their FESEM. We would also like to thank the Braunschweig International Graduate School of Metrology (IGSM) for their support.

References

1. Li, S. F. and Waag, A. (2012). GaN based nanorods for solid state lighting, Appl. Phys. Rev., 111, pp. 071101-1 - 071101-23.

2. Waag, A., Wang, X., Fündling, S., Ledig, J., Erenburg, M., Neumann, R., Al-Suleiman, M. A. M., Wei, J. D., Li, S. F., Wehmann, H.-H., Bergbauer, W., Straßburg, M., Trampert, A. and Riechert, H. (2011). The nanorod approach: GaN NanoLEDs for solid state lighting, Phys. Stat. Sol. (c), 7-8, pp. 2296-2301.

3. Schwarz, U. T. and Kneissl, M. (2007). Nitride emitters go nonpolar, Phys. Stat. Sol. RRL, 1, No.3, A44-A46.

4. Kim, H.-M., Cho, Y.-H, Lee, H., Kim, S. II, Ryu, S. R., Kim, D. Y., Kang, T. W., Chung, K. S. (2004). High-brightness light emitting diodes using dislocation-free indium gallium nitride/gallium nitride multiquantum-well nanorod arrays, Nano Lett., 4, pp. 1059-1062.

5. Dadgar, A., Bläsing, J., Diez, A., Alam, A., Heuken, M. and Krost, A. (2000). Metalorganic Chemical Vapor Phase Epitaxy of Crack-Free GaN on Si (111) Exceeding 1 μm in Thickness, Jpn. J. Appl. Phys., 39, pp. L1183-L1185.

6. Reiher, A., Bläsing, J., Dadgar, A., Diez, A. and Krost, A. (2003). Efficient stress relief in GaN heteroepitaxy on Si(111) using low-temperature AlN interlayers, J. Cryst. Growth, 248, pp. 563-567.

7. Dadgar, A., Hums, C., Krost, A., Diez, A., Schulze, F. and Bläsing, J. (2006). Epitaxy of GaN LEDs on large substrates: Si or sapphire?, Proc. SPIE, 6355, pp. 63550R-1 - 63550R-08.

8. www.osram-os.com

9. Bengoechea-Encabo, A., Barbagini, F., Fernandez-Garrido, S., Grandal, J., Ristic, J., Sanchez-Garcia, M. A., Calleja, E., Jahn, U., Luna, E. and Trampert, A. (2011). Understanding the selective area growth of GaN nanocolumns by MBE using Ti nanomasks, J. Cryst. Growth, 325, pp. 89-92.

10. Kouno, T., Kishino, K., Yamano, K. and Kikuchi, A. (2009). Two-dimensional light confinement in periodic InGaN/GaN nanocolumn arrays and optically pumped blue stimulated emission, Optics Express, 17, pp. 20440-20447.

11. Consonni, V., Knelangen, M., Geelhaar, L., Trampert, A. and Riechert, H. (2010). Nucleation mechanisms of epitaxial GaN nanowires: Origin of their self-induced formation and initial radius, Phys. Review B, 81, pp. 085310-1 - 085310-10.

12. Hersee, S. D., Sun, X. and Wang, X. (2006). The controlled growth of GaN nanowires, Nano Lett., 6, pp. 1808-1811.

13. Hong, Y. J., Lee, C.-H., Yoon, A., Kim, M., Seong, H.-K., Chung, H. J., Sone, C., Park, Y. J. and Yi, G.-C. (2011). Visible-Color-Tunable Light-Emitting Diodes, Adv. Mater., 23, pp. 3284-3288.

14. Verma, J., Simon, J., Protasenko, V., Kosel, T., Xing, H. G. and Jena, D. (2009). N-polar III-nitride quantum well light-emitting diodes with polarization-induced doping, Appl. Phys. Lett., 99, pp. 171104-1 - 171104-3.

15. Rajan, S., Chini, A., Wong, M. H., Speck, J. S. and Mishra, U. K. (2007). N-polar GaN/AlGaN/GaN high electron mobility transistors, J. Appl. Phys., 102, pp. 044501-1 - 044501-6.

16. Li, Z. Q., Lestrade, M., Xiao, Y. G. and Li, S. (2011). Effects of polarization charge on the photovoltaic properties of InGaN solar cells, Phys. Stat. Sol. (a), 208, pp. 928-931.

17. Li, S. F., Fuendling, S., Wang, X., Merzsch, S., Al-Suleiman, M. A. M., Wei, J. D., Wehmann, H.-H., Waag, A., Bergbauer, W. and Strassburg, M., (2011). Polarity and its influence on growth mechanism during MOVPE growth of GaN sub-micrometer rods, Cryst. Growth Des., 11, pp. 1573-1577.
18. Chen, X. J., Perillat-Merceroz, G., Sam-Giao, D., Durand, C. and Eymery, J. (2010). Homoepitaxial growth of catalyst-free GaN wires on N-polar substrates, Appl. Phys. Lett., 97, pp. 151909-1 - 151909-3.
19. Bergbauer, W., Strassburg, M., Kölper, Ch., Linder, N., Roder, C., Lähnemann, J., Trampert, A., Fündling, S., Li, S. F., Wehmann H.-H. and Waag, A. (2011). N-face GaN nanorods: Continuous-flux MOVPE growth and morphological properties, J. Cryst. Growth, 315, pp. 164-167.
20. Lu, H., Cao, D. S., Xiu, X. Q., Xie, Z. L., Zhang, R., Zheng, Y. D., Li, Z. H. (2008). Schottky rectifiers fabricated on bulk GaN substrate analyzed by electron-beam induced current technique, Solid-State Electronics, 52, pp. 817-823.
21. Fündling, S., Sökmen, Ü., Peiner, E., Weimann, T., Hinze, P., Jahn, U., Trampert, A., Riechert, H., Bakin, A., Wehmann, H.-H. and Waag, A. (2008). Gallium-nitride heterostructures on 3D structured silicon, Nanotechnology, 19, pp. 405301-1 - 405301-6.
22. Bae, S.-Y., Kim, D.-H., Lee, D.-S., Lee, S.-J. and Beak, J. H. (2011). Pentacene-Based Low-Leakage Memory Transistor Dielectric/Electrolytic/Dielectric Polymer Layers, Electrochem. Solid-State Lett., 15, pp. H47-H50.
23. Wildeson, I. H., Colby, R., Ewoldt, D. A., Liang, Z., Zakharov, D. N., Zaluzec, N. J., García, R. E., Stach, E. A. and Sand, T. D. (2010). III-nitride nanopyramid light emitting diodes grown by organometallic vapour phase epitaxy, J. Appl. Phys., 108, pp. 044303-1 - 044303-8.
24. Scholz, F. (2012). Semipolar GaN grown on foreign substrates: a review, Semicond. Sci. Technol., 27, pp. 024002-01 - 024002-15.
25. Bergbauer, W., Strassburg, M., Kölper, Ch., Linder, N., Roder, C., Lähnemann, J., Trampert, A., Fündling, S., Li, S. F., Wehmann, H.-H. and Waag, A. (2010). Continuous-flux MOVPE growth of position-controlled N-face GaN nanorods and embedded InGaN quantum wells, Nanotechnology, 21, pp. 305201-1 - 305201-5.
26. Liu, F., Collazo, R., Mita, S., Sitar, Z., Duscher, G. and Pennycook, S. J. (2007). The mechanism for polarity inversion of GaN via a thin AlN layer: Direct experimental evidence, Appl. Phys. Lett., 91, pp. 203115-01 - 203115-03.
27. Koester, R., Hwang, J. S., Durand, C., Dang, D. Le Si and Eymery, J. (2010). Self-assembled growth of catalyst-free GaN wires by metal–organic vapour phase epitaxy, Nanotechnology, 21, pp. 015602-01 - 015602-09.
28. Hiramatsu, K., Nishiyama, K., Motogaito, A., Miyake, H., Iyechika, Y. and Maeda, T. (1999). Recent Progress in Selective Area Growth and Epitaxial Lateral Overgrowth of III-Nitrides: Effects of Reactor Pressure in MOVPE Growth, Phys. Stat. Sol. (a), 176, pp. 535-543.
29. Wei, J. D., Neumann, R., Wang, X., Li, S.F., Fündling, S., Merzsch, S., Al-Suleiman, M.A.M., Sökmen, Ü., Wehmann, H.-H. and Waag, A. (2011). Polarity analysis of GaN nanorods by photo-assisted Kelvin probe force microscopy, Phys. Stat. Sol. C, 8, pp. 2157.
30. VanMil, B. L., Guo, H. C., Holbert, L. J., Lee, K.-N., Myers, T. H., Liu, T. and Korakakis, D. (2004). High temperature limitations for GaN growth by rf-plasma assisted molecular beam epitaxy: Effects of active nitrogen species, surface polarity, hydrogen, and excess Ga-overpressure, J. Vac. Sci. Technol. B, 22, pp. 2149-2154.
31. Northrup, J. E. and Neugebauer, J. (2004). Strong affinity of hydrogen for the GaN (000-1) surface: Implications for molecular beam epitaxy and metalorganic chemical vapor deposition, Appl. Phys. Lett., 85, pp. 3429-3431.
32. Feenstra, R. M., Dong, Y., Lee, C. D. and Northrup, J. E. (2005). Recent Developments in Surface Studies of GaN and AlN, J. Vac. Sci. Technol. B, 23, pp. 1174-1180.

33. Wulff, G. (1901). Zur Frage der Geschwindigkeit des Wachstums und der Auflösung von Krystallflächen, Z. Kristallogr. Mineral., 34, pp. 499-530.
34. Hiramatsu, K., Nishiyama, K., Onishi, M., Mizutani, H., Narukawa, M., Motogaito, A., Miyake, H., Iyechika, Y. and Maeda, T. (2000). Fabrication and characterization of low defect density GaN using facet-controlled epitaxial lateral overgrowth (FACELO), J. Cryst. Growth, 221, pp. 316-326.
35. Li, S.F., Wang, X., Fündling, S., Al-Suleiman, M. A. M., Erenburg, M., Wei, J. D., Wehmann, H.-H., Waag, A., Mandl, M., Bergbauer, W. and Strassburg, M. (2011). Dependence of N-polar GaN nanorods morphology on growth parameters during selective area growth by MOVPE, Proceedings of XIV European Workshop on Metalorganic Vapor Phase Epitaxy (EW-MOVPE), pp. D18 (Wroclaw, Poland).
36. Ng, H. M., Weimann, N. G. and Chowdhury, A. (2003). GaN nanotip pyramids formed by anisotropic etching, J. Appl. Phys., 94, pp. 650-653.
37. Wei, J. D., Li, S. F., Atamuratov, A., Wehmann, H.-H. and Waag, A. (2010). Photoassisted Kelvin probe force microscopy at GaN surfaces: The role of polarity, Appl. Phys. Lett., 97, pp. 172111-1 - 172111-3.
38. Hiramatsu, K., Nishiyama, K., Motogaito, A., Miyake, H., Iyechika, Y. and Maeda, T. (1999). Recent Progress in Selective Area Growth and Epitaxial Lateral Overgrowth of III-Nitrides Effects of Reactor Pressure in MOVPE Growth, Phys. Stat. Sol. (a), 176, pp. 535-543.
39. Reshchikov, M. A. and Morkoc, H. (2005). Luminescence properties of defects in GaN, J. Appl. Phys., 97, pp. 061301-01 - 061301-95.
40. Mierry, P. De, Ambacher, O., Kratzer, H. and Stutzmann, M. (1996). Yellow Luminescence and Hydrocarbon Contamination in MOVPE-Grown GaN, Phys. Stat. Sol. (a), 158, pp. 587-597.
41. Sun, Q., Cho, Y. S., Lee, I.-H., Han, J., Kong H. and Cho, H. K. (2008). Nitrogen-polar GaN growth evolution on c-plane sapphire, Appl. Phys. Lett., 93, pp. 131912-1 - 131912-3.
42. Du, D. C., Zhang, J. C., Ou, X. X., Wang, H., Chen, K., Xue, J. S., Xu, S. R. and Hao, Y. (2011). Investigation of yellow luminescence intensity of N-polar unintentionally doped GaN, Chin. Phys. B, 20, pp. 037805.
43. Zywietz, T., Neugebauer, J. and Scheffler, M. (1998). Adatom diffusion at GaN (0001) and (000-1) surfaces, Appl. Phys. Lett., 73, pp. 487-489.
44. Sumiya, M., Yoshimura, K., Ito, T., Ohtsuka, K. and Fuke, S. (2000). Growth mode and surface morphology of a GaN film deposited along the N-face polar direction on c-plane sapphire substrate, J. Appl. Phys., 88, pp. 1158-1165.

PROGRESS IN SiC MATERIALS/DEVICES AND THEIR COMPETITION

DIETER K. SCHRODER

School of Electrical, Computer and Energy Engineering, Arizona State University
Tempe, AZ 85287-5706, USA
schroder@asu.edu

Power semiconductor devices are important for numerous applications with power conversion being an important one. Wide energy gap semiconductors SiC and GaN have properties that make them attractive for such applications. Among these properties are high thermal conductivity, high breakdown electric field, wide energy gap, low intrinsic carrier concentration, high thermal stability, high saturation velocity and chemical inertness. These lead to low *on*-resistance, high breakdown voltage, high frequencies, small volume, and small passive inductors and capacitors. These desirable properties are offset by the higher material costs and higher defect densities. Although wide energy gap devices have been in development for many years, only recently have they become available commercially. Their main competition is silicon power devices with breakdown voltages up to 8000 V and very high surge current capacity. However, silicon power devices are approaching their material limits and wide energy gap devices are beginning to have an impact in the power electronics space. SiC has the advantage of substrates with diameters approaching 150 mm and the ability to grow thermal SiO_2. GaN has the heterojunction advantage, but no viable substrate technology. In fact, a large portion of SiC production is used for GaN substrates. GaN material development has also benefited significantly from the development of optical devices, *e.g.,* light-emitting diodes and lasers.

Keywords: Semiconductor; silicon carbide; silicon; gallium nitride; power device; MOSFET; super MOSFET; insulated-gate bipolar transistor; IGBT; Schottky diode; junction FET; *on*-resistance; breakdown voltage.

1. Introduction

Silicon Carbide is a very hard and inert semiconductor material which has been used in abrasive products such as grinding wheels for more than one hundred years. It was already used in the early 1900s as diode detectors in radios[1] and the first light-emitting diodes were made of SiC.[2] By 1995, commercial wafers of reasonable size and quality were grown for the electronics industry. In the summer of that year, the jewelers at Charles and Colvard realized the potential of this material as gemstones which were claimed to have superior properties to other gemstones (index of refraction ~2.76 which is larger than 2.42 for diamond).[3] They named this jewel *moissanite* after Henri Moissan, who discovered natural SiC in 1893 left by a meteor in Diablo Canyon (Meteor Crater) in Arizona.[4] SiC also has many desirable properties for semiconductor device applications. Naturally occurring SiC is found only in minute quantities in certain types of meteorite, in corundum deposits and in kimberlite. Virtually all SiC, including moissanite jewels, is synthetic. However, SiC crystal quality, in spite of significant improvements, is not yet adequate for large-scale power device production.

William Shockley wrote in 1959 at the *First International Conference on Silicon Carbide.*[5]

Today, in the electronics field there are probably two areas of special interest. One of these areas is miniaturization, the process of making devices small, complicated and fast; the other area has to do with problems of new environment, such as higher temperatures and radiation resistance. It is to this area of high temperature, of course, that this Conference is particularly relevant. Now, the big question is this: How is the problem of high temperature going to be solved? What are the horses to put one's money on?

Speaking in terms of the impressions of other people, because this is not a field in which I myself have worked at all extensively, there are two approaches. One approach is the logical sequence we see here: Ge, Si, SiC, C, i.e., the use of Ge-Si-C ternary materials. This is obviously one course of action. The other one is the III-V Compounds or possibly others like them. In facing a choice here, one is on the horns of a dilemma, because neither situation is particularly bright. The SiC situation suffers from the very same thing that makes it good. The bond is very strong and so all processes go on at a very high temperature. In the case of III-V materials, one has an unlimited supply of donor and acceptor impurities in the sense that any lack of stoichiometry produces conductivity.

Since I'm now moving towards the state of becoming an elder statesman in this field, I would like to reminisce on some of the SiC problems with respect to the early days of Si and Ge. Well, the situation is obviously much worse than the Si situation. There is, however, one difference. Today, we have much better theoretical and experimental tools to work with. Another aspect of the silicon carbide situation is similar to past situations in the semiconductor field. The lesson is that one should not give up too soon and one should not always look for gold at the ends of new rainbows. In 1945, germanium was probably the best understood semiconductor. However, at Bell Telephone Labs we said that ionization energies had not been measured; impurity band conduction was sort of a gleam in somebody's eye and degeneracy in the energy bands was just beginning to be appreciated. We decided that this situation would have to be improved by some accurate measurements. We thought that within a few years one would measure the ionization energies, have some better Hall Effect measurements and so on. The energy bands were certainly spherical because this was all fitting in rather nicely. Then along came better crystals.

Now in regard to these crystals I would like to tell this story. It is a story about myself. I did not feel at that time it was really very important to try to make big crystals. I felt that one could obtain sufficiently large single crystals from the grains of large polycrystalline ingots. It was the device people, people who wanted to make uniform transistors, who really put more emphasis on obtaining big crystals. The strongest proponents were Morton and his associates at Bell Telephone Laboratories. Of course, when the big, uniform crystals were obtained, all kinds of physics started growing out of them. One of the reasons for this was that the phenomena one was working with, such as diffusion lengths, began to get bigger than the distance one could move micromanipulators. So one could get inside the scale of the phenomena, and this seems to me to be just what you are barely on the edge of being able to do with silicon carbide. Silicon followed germanium and was more difficult. Indeed, not all problems for silicon have been solved as yet. One may be able to get a piece of pure, zone refined, silicon; however, this crystal will contain many dislocations. On the other hand, one may be able to grow dislocation-free single crystals of silicon, but then there is, again, the oxygen-impurity problem. This problem, I think, may soon be solved. The situation may be similar with silicon carbide. The material problem will have to be extensively worked on. Perhaps one day somebody will find a suitable solvent from which large single crystals of silicon carbide will be grown easily. On the other hand, the approach might very well be different. Perhaps something on a really large scale has to be done, so that large crystals will result.

Wide energy gap semiconductors have several significant advantages over more conventional, *e.g.*, Si, semiconductors as their material properties (high thermal conductivity, high breakdown electric field, wide energy gap, low intrinsic carrier concentration, high thermal stability, high saturation velocity and chemical inertness) make them suitable for high-power and high-frequency devices.[6] The increased operating temperature and high thermal conductivity of SiC reduces the weight, volume, cost, and complexity of thermal management systems. The high-temperature behavior is important for applications including automotive, aerospace, deep-well drilling, and other industrial systems. Figure 1 (a) shows major applications for high-temperature electronics (HTE). Many of these applications are in the range up to about 300°C, a temperature barely accessible by silicon devices. While Si devices have operated briefly up to 400°C, reliability is a major concern. Even though the physics of SiC allows for maximum junction temperatures well above 200°C, limitations in interconnects and packaging technology and the lack of other parts at the system level able to work at this temperature pose a serious obstacle for the utilization of the inherent high-temperature potential of SiC.[7] Higher breakdown electric fields place greater demands on package insulators.

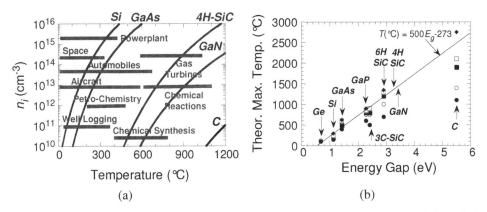

Fig. 1. (a) Intrinsic carrier density and typical operating temperatures versus temperature, (b) theoretical maximum temperatures versus energy gap.[8]

Oil and gas well are drilled deeper with rising bottom hole temperatures. Hot oil/gas wells have temperatures up to 200°C and geothermal wells operate near 350°C. To increase the useful life and production capacity of multilateral wells, the need for reliable

HTE increases to monitor the pressure, temperature, and multiphase flow for optimized production, requiring permanently installed gauges and data acquisition systems to last for up to 10 years in high-temperature environments, well beyond the capabilities of Si electronics. High-temperature electronics are also ideal for distributed aircraft and ground based turbine engine control and for distributed heavy duty engine controls in large diesel or internal combustion engines, including cylinder pressure transducer output amplifiers and bus communication electronics, exhaust gas recovery valve control electronics and engine control electronics that have to operate above 150°C.

Automotive electronic modules are designed for a maximum operating temperature of 125°C. The temperatures under the automobile hood are likely to increase above today's levels due to factors such as smaller radiator and lower radiator openings and greater engine compartment "fill factor", reducing the amount of air flowing over the engine. Furthermore, electronic components mounted directly on the engine and transmission and within actuators reduce cost and simplify assembly. This is especially important for wheel-mounted components where the temperature can reach 250°C. Along exhaust systems and for cylinder-specific monitoring, temperatures reach 400-800°C.

An important semiconductor material parameter for HTE is the intrinsic carrier density, n_i. The limits imposed by n_i are illustrated in Fig. 1(a). When n_i reaches the device's doping concentration, devices cease to function properly, *e.g.*, in *pn* junctions the barriers disappear and they become linear. Assuming a doping concentration of 10^{15}-10^{16} cm^{-3}, we see that Si has an upper temperature limit of about 300-400°C. Some SOI devices (opamps, analog switches, and microcontrollers) are specified to 225°C and usable up to 275-300°C.[9] GaAs extends this limit to about 300-400°C and beyond that we must look to the wide energy gap semiconductors SiC, GaN, and diamond. Typical temperature ranges for various high-temperature applications are also shown on Fig. 1(a). When the temperature of the environment is too high, silicon-based microelectronic integrated circuits to monitor and/or control crucial hot-section subsystems must reside in cooler areas that are either remotely located or actively cooled with air or liquid cooling medium pumped in from elsewhere. Such thermal management approaches introduce additional overhead that can negatively offset the desired benefits of the electronics relative to overall system operation. The additional overhead, in the form of longer wires, extra connectors, and/or cooling system plumbing, can add undesired size and weight to the system, as well as increased complexity, part count, and corresponding increased potential for failure. These difficulties stand as major hindrances towards expanded use of electronics to improve hot-environment system performance.

Despite these difficulties, the drive to put even more electronics into various applications, including subsystems that monitor and control high temperature areas of aircraft and automobiles, is unlikely to slow in the near future. A mere handful of high temperature electronics chips, perhaps purchased for a few hundred dollars, can enable millions of dollars of increased capability to a very large system. For example, directional and compositional telemetry made possible by ~200°C high temperature electronics in a deep-well drilling operation can help prevent 10s to 100s of millions of dollars of loss in equipment and resources.[10] Similarly, improved weight, fuel economy, and maintenance over the multi-decade life of a commercial passenger aircraft would also translate into substantial operating cost savings. The higher efficiency and higher switching frequencies lead to lower cooling requirements and smaller heat sinks. Furthermore, the size and cost of passive components, mainly inductances, can be reduced with higher switching frequency and maintain high efficiency, important for inverters, electric vehicles and switched power supplies.

Most research has focused on current–voltage and gain-frequency behavior with little mention of how long such parts operate at high temperature. However, most envisioned applications require reliable operation over long times at high temperature, on the order of thousands of hours or more. Without such long-term durability, extreme temperature semiconductor ICs will not practically benefit and will not be inserted into the vast majority of important intended applications, including long-term stability of insulators in MOS devices and Schottky barrier contacts. Devices using highly stable SiC pn junctions (BJTs) and JFETs, offer the best chance for durable extreme temperature operation.[11] High material cost and material effects that lead to device degradation are another impediment to large scale commercial introduction. A comparison of SiC properties with Si is shown in the spider diagram in Fig. 2.[12] The energy gap, breakdown electric field, dielectric constant, saturation velocity and thermal conductivity are normalized to unity for Si. SiC outperforms Si in all regions of this figure.

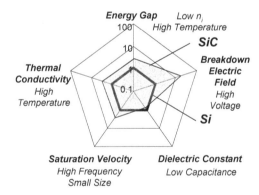

Fig. 2. Material properties of SiC compared to Si. The "1" refers to Si.

2. Materials

SiC grows in numerous stacking orders (~250 polytypes[13]), each with its own energy gap. All polytypes have a hexagonal frame with a carbon atom situated above the center of a triangle of Si atoms and below a Si atom belonging to the next layer. The difference between the polytypes is the stacking order between succeeding double layers of carbon and silicon atoms. The most common polytypes are cubic 3C ($E_G = 2.36$ eV), hexagonal 6H ($E_G = 3.0$ eV) and hexagonal 4H ($E_G = 3.26$ eV). The different polytypes differ only in the stacking of double layers of Si and C atoms and this affects all electronic and optical properties of the crystal. The 4H polytype is favored over the 6H in high-power vertical devices because of its higher mobility along the *c*-axis - the growth direction for epitaxial layers. SiC wafers, generally grown by the physical vapor transport method using the Lely process[14] on basal (0001) plane seeds, are available with 75-100 mm diameter. Typical defect densities are: micropipes: 0-5 cm^{-2}, dislocations: 10^3-10^4 cm^{-2}. Epitaxial layers, grown by chemical vapor deposition, typically contain 1-100 cm^{-2} dislocations and may contain other defects such inclusions and stacking faults.

The (0001) Si-terminated face has the slowest growth rate exhibiting the best stability against temperature and concentration fluctuations. There are four hexagonal Miller indices describing the directions in all SiC poly-types except for the 3C poly-type where the normal cubic notation is used. The last index refers to the *c*-direction and the three first describe directions in the basal plane. In Fig. 3 the stacking sequence is shown for 3C, 4H and 6H. If the first double layer is called the A position, the next layer that can be placed according to a closed packed structure will be placed on the B position or the C position. The different poly-types will be constructed by permutations of these three positions. The number denotes the periodicity and the letter the resulting structure which in this case is hexagonal. The 3C-SiC poly-type is the only cubic poly-type and it has a stacking sequence ABCABC... or ACBACB.[15]

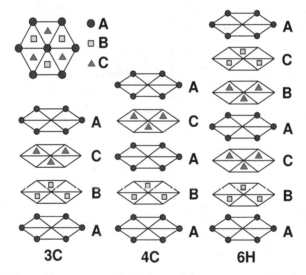

Fig. 3. The stacking sequence of double layers of the most common SiC polytypes.

2.1. *Bulk Defects*

Occasionally, there are 3C poly-type inclusions (double stacking faults) in 4H epi layers that often originate at the epi/substrate interface with a pit or particle as the starting point. These inclusions form by a slight re-arrangement of the atoms during growth. Figure 4(a) shows a cross-sectional transmission electron microscopy (TEM) image taken just below the surface of a 4H-SiC film.[16] Three thin stripes perpendicular to the film growth direction [0001] can be seen embedded within the normal 4H-SiC lattice. In the higher resolution lattice image Fig. 4(b), the stripes are identified as layers of 3C poly-type with 13 bilayers of Si-C resulting in 4H/3C/4H-SiC quantum well structures. Such inclusions act as quantum wells in the 4H matrix (Fig. 4(c)), since they have a lower energy gap. Most of the band offset is in the conduction band and they have a strong electric field due to the spontaneous polarization in the 4H material, so they are trapezoidal, not rectangular in shape. When they intersect the surface they cause local reductions in Schottky barrier height since the 3C energy gap is lower than that of 4H.

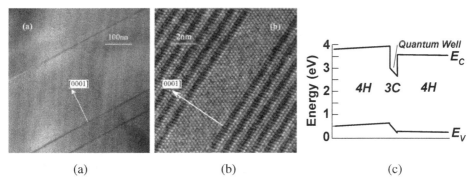

(a)   (b)   (c)

Fig. 4. (a) Bright-field TEM image (b) high-resolution lattice image of the cross section of a 4H-SiC film with 3C inclusions and (c) band diagram of one 3C quantum well.

2.2. *Carrier Lifetimes*

The minority carrier lifetime in SiC is usually quite low, typically in the low μs region or below in spite of the indirect energy gap. While lifetimes play essentially no role in unipolar MOSFETs and Schottky barrier diodes, they are very important in bipolar devices (*pn* diodes, IGBTs, BJTs) because they provide low *on* resistances due to conductivity modulation. In contrast to Si, where metallic impurities are the major cause of lifetime reduction, in SiC structural defects appear to be the main culprit. Among these are grain boundaries, $Z_{1/2}$ (E_C-0.65 eV) and/or EH6/7 (E_C-1.55 eV) defects each thought to be defects on cubic and hexagonal sites, most likely carbon vacancy related. The $Z_{1/2}$ defect, whose density is independent of growth rate and dependent primarily on C/Si ratio and growth temperature, is the most lifetime-limiting defect.[17] An almost linear correlation between these two defects has been established for $Z_{1/2}$ concentrations above 10^{13} cm^{-3}.[18]

The inverse lifetime is plotted in Fig. 5 versus $Z_{1/2}$ concentration, exhibiting a linear relationship down to about 2×10^{13} cm^{-3}. Below that concentration, the lifetime becomes constant and is dominated by "other" defects. The transition concentration of 10^{13} cm^{-3} is, of course, dependent on surface recombination and the density of "other" defects, both of which are likely to vary from sample to sample. Measured lifetimes frequently exhibit wide variations between researchers. This should not be surprising, as lifetimes

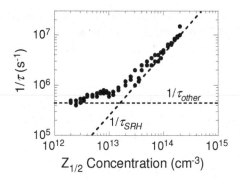

Fig. 5. Relation between the inverse carrier lifetime and the $Z_{1/2}$ concentration measured with an illumination intensity of 2×10^{14} cm^{-2}. 4*H*-SiC epilayer thickness of 44–53 μm.[19]

depend very much on the sample and the measurement (surface condition, injection level, temperature, etc.) and technique (microwave photoconductance decay, photoluminescence decay, free carrier absorption, etc.).

2.3. *Oxide and Interface Traps*

Even though SiC MOSFETs became recently commercially available, they still have stability problems. Furthermore, the effective mobility is quite low for a variety of reasons. Among these are the high interface trap densities (D_{it}) at the SiO$_2$/SiC interface. Various types of interface traps are believed to exist in SiO$_2$/SiC systems. Pb centers probably exist and are likely similar to those in the SiO$_2$/Si system, but play only a marginal role for SiC. Excess carbon at the interface has been observed and is believed to be one source of interface traps through Si-C-O bonded interlayer leading to interface traps with energy levels throughout the energy gap. Near-interface traps (NIT) near the conduction band of 4H-SiC have been observed although their origin is not known. They may be due to Si-Si pairs with Si driven into the oxide by stress between SiO$_2$ and SiC. The interface trap density is schematically illustrated in Fig. 6.[20]

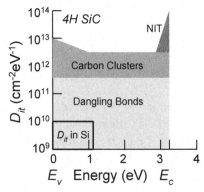

Fig. 6. SiO$_2$/SiC interface trap densities. D_{it} is composed of dangling bonds, carbon clusters and near-interface traps.

Interface trap measurements are difficult to interpret due to the wide energy gap of SiC. For example, electron/hole emission from states near the center of the energy gap takes $\sim 10^{16}$ s at room temperature! However, since carrier emission from states near the middle of the energy gap takes $\sim 10^{16}$ s, these traps are never in near-equilibrium, and D_{it} is underestimated. If we limit ourselves to lower measurement frequencies around 1 Hz, D_{it} to only ~ 0.5 eV from the band edge can be determined at room temperature. Even methods that do not rely on ac measurements, *e.g.*, the Terman method which measures the shift of the *C-V* curve from its ideal value, usually underestimate the interface trap density.[21] D_{it} is higher on the C face than on the Si face. However, the SiO_2/SiC interface is still poorly understood and only limited information about the atomistic structure of the interface exists. It is generally accepted that the high near-band edge D_{it} accounts for the low effective mobilities of SiC MOSFETs. The main approach to reduce D_{it} has been annealing in a nitrogen-containing ambient at or near the oxidation temperatures of 1100-1200°C,[22] leading to nitrogen incorporation within 1-2 nm of the SiO_2/SiC interface.[23] Recently, phosphorus has also been used to passivate the interface, with trap density and effective *n*-channel mobility approximately 2 times lower and higher, respectively, compared to nitrogen.[24]

Hydrogen annealing, effective in Si MOS technology, generally has little beneficial effect, likely due to a more complex nature than singular dangling bonds.[25] However, monatomic hydrogen reduces D_{it}.[26] It has been proposed that H passivates three-fold coordinated C and Si atoms in the bonded Si-C-O interfacial region. Nitrogen, on the other hand, is an effective passivant for C and Si dangling bonds. Phosphorus may be more efficient in passivating traps similar to those passivated by nitrogen, and there is currently discussion whether nitrogen and phosphorus may counter-dope the SiC surface to produce a buried channel effect. The effects of nitrogen, hydrogen and phosphorus anneals on the field-effect mobility of (0001) *n*-4H-MOSFETs are shown in Fig. 7(a). The mobility is most enhanced for phosphorus, while a H_2(Pt) anneal after an nitric oxide (NO) anneal increases the mobility by 20-30% compared to NO alone. The increase in mobility is due to passivation of acceptor-like interface traps in the upper half of the 4H-SiC energy gap (Fig. 7(b)). Nitrogen and phosphorus effectively reduce the very high trap density near the conduction band edge, while hydrogen is more effective for the acceptor-like states near mid-gap.

Bias-temperature instability (BTI) is a significant reliability concern in Si MOS devices.[27] Only a few papers dealing with BTI in SiC have been published,[28,29] but BTI will almost certainly be problematic in SiC MOS technology. The ensuing D_{it} increase due to BTI is especially detrimental to MOSFET mobility. Steady-state and pulsed capacitance and conductance measurements of SiC MOS capacitors give evidence of BTI. The normalized data of four consecutive *C-t* measurements are shown in Fig. 8(a). Each curve recovers faster than the previous one, indicating reduced lifetime after each pulsed MOS-C measurement. The MOS-C recovery time depends on interface and bulk space-charge region electron-hole pair (*ehp*) generation. We do not believe that these pulsed measurements alter the bulk generation rate, but rather enhance interface generation.

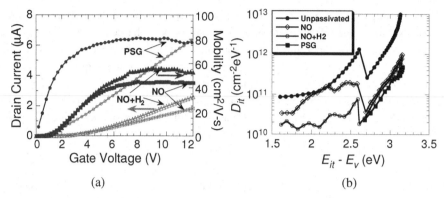

(a)   (b)

Fig. 7. (a) I_D-V_G and μ_{FE} for n-MOSFETs after NO, NO+H$_2$ and phosphorous passivation. Without passivation, μ_{FE} < 10 cm^2/V-s; (b) D_{it} for thermal SiO$_2$ on (0001) n-4H-SiC. NO – 1175°C, 2hr, 1atm, pure NO.   H$_2$: 1hr, 500°C, 1atm, pure H$_2$ following an NO anneal and Pt gate deposition. Phosphorous: 6hr, 1000°C with a planar diffusion source (SiP^{3}O^7 with N$_2$ carrier gas). After J. Williams, Auburn University.

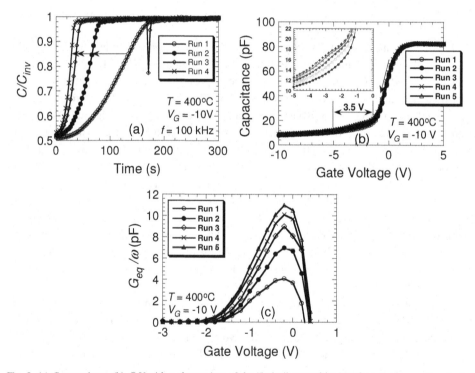

Fig. 8. (a) C-t transients, (b) C-V with a closer view of the "bulge" created by interface state generation in the inset, and (c) the corresponding equivalent parallel conductance.

During NBTI stress in the SiO$_2$/Si system, D_{it} increases with time, leading to increased surface generation and decreased effective generation lifetimes. The steady-state C-V and G-V data in Fig. 8(b) show that the interface trap density has increased after each pulsed MOS measurement. The C-V curves show stretch-out in the deep-depletion regime, attributed to higher D_{it}. This interface "stretch-out" becomes more significant with each measurement, indicative of interface trap generation as a result of the electrical stress. The interface trap density is estimated as $N_{it} \approx \Delta V_{FB} C_{ox}/q = 1.6 \times 10^{12}$ cm^{-2}. The equivalent parallel conductance which is more sensitive than the capacitance to changes in interface trap density, exhibit a significant increase in the conductance as shown by the G_{eq}/ω-V data in Fig. 8(c).

Carrier injection due to gate bias voltage is also of concern. Figure 9(a) shows room-temperature Fowler-Nordheim data of n-4H MOS capacitors biased in inversion. The NO-passivated capacitors clearly show evidence of enhanced charge trapping with increasing hole injection, most likely associated with nitrogen at the O-S interface, since no charge trapping is observed for unpassivated capacitors. Interfacial nitrogen actually suppresses electron injection from the SiC into the oxide[30] so this behavior for holes is quite different, though not unlike similar behavior observed for SiO$_2$/Si. Based on results of Fig. 9(b), the trapped charge is believed to be due to interface trap generation and hole trapping. The h-f C-V curves shift left and stretch out under room temperature F-N stress. The stretch-out indicates interface trap generation, and the shift to more negative V_G indicates the injection and trapping of positive charge. The hole traps may result from three-fold coordinated nitrogen with a singly occupied lone pair state in the lower part of the SiC energy gap.[31] This state could be associated with 'extra' nitrogen atoms incorporated in the oxide and not necessarily involved in the passivation of acceptor-like traps in the upper half of the SiC energy gap.

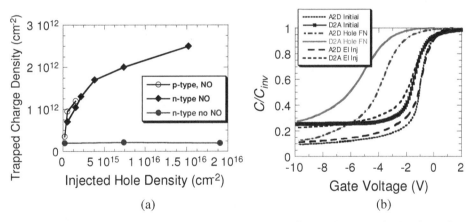

(a)   (b)

Fig. 9. (a) Trapped charge following F-N hole injection for n-4H-SiC MOS-Cs w/ and w/o NO nitridation. Results for NO-annealed p-4H-MOS capacitor in accumulation shown for comparison, (b) hf C-V curves for NO annealed n-4H-SiC MOS-Cs. After the A to D sweeps with the device in depletion, UV light creates a p-inversion layer for hole injection. Stretch-out and left shift in the red curves indicate interface trap generation and hole trapping. After J. Williams, Auburn University.

3. SiC Devices

In spite of the fact that SiC has been researched for over 50 years, there is a dearth of commercial devices. The first commercial devices were Schottky diodes followed by junction FETs and more recently MOSFETs. Wide energy gap semiconductor devices will not replace Si for integrated circuits. Their main application for the foreseeable future, aside from optical applications, will be in power management where their higher temperature and breakdown voltage behavior has definite advantages over silicon. Both SiC and GaN will play a role for this application. GaN, with an energy gap similar to SiC, takes advantage of its heterojunction devices and substantial development investment for light-emitting devices. One disadvantage is the absence of bulk material, requiring foreign substrates, *e.g.*, SiC. SiC is grown in ingots with diameters to 100 mm (150 mm in the near future) and is has the distinct advantage of thermally-grown SiO_2 allowing MOS devices. However, the SiO_2/SiC interface still exhibits stability problems due to charge injection and high interface trap densities. At present, SiC MOSFETs are available for voltages of ~1000 V and higher, whereas GaN devices are currently at 200 V and below. Several companies have recently announced 600 V GaN devices. Silicon power MOSFETs are approaching their theoretical limits for *on* resistance and gate charge.

The main applications of SiC devices are in power management. For $V_{BD} < 1000$ V, the chief applications are: switch-mode power supplies (lower power density, higher frequency allows smaller and lighter modules; higher efficiency), uninterrupted power supply (UPS), power factor correction and others. For $V_{BD} > 1000$ V, the chief applications are hybrid electric vehicles, motor drives, avionics (lower weight), inverters/converters for alternative energy (solar, wind), defense (Navy: electric ships). Recent SiC inverter with ¼ the volume of a Si inverter with 70% reduced power loss has been introduced.

The size of a power supply's components determines its power density; the largest components are the inductors and capacitors with MOSFETs a distant third. Capacitors and inductors shrink with higher switching frequency, which is topping out for Si MOSFETs. Switching losses due to the gate charge cancel any further increase in switching speed. Shrinking the die of Si MOSFETs unfortunately increases R_{on}. MOSFETs have traditionally measured their performance figure-of-merit as R_{on} times the gate charge. Decreasing R_{on} usually comes from increasing the device area, but that increases the gate charge and the capacitance, decreasing the device's speed.

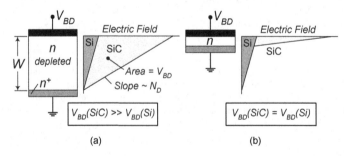

Fig. 10. Schottky diode cross sections and ε-x plots for (a) long- and (b) short-based diodes showing the effect of high critical electric field.

The device cross sections and electric field-distance plots of two reverse-biased Schottky diodes are shown in Fig. 10. The basic equations to explain these devices are

$$\frac{d\varepsilon}{dx} \sim N_D; \ V_{BD} \sim \int \varepsilon_{crit} dx; \ R_{on} = \frac{\rho W}{A} = \frac{W}{q\mu_n N_D A} \sim \frac{W}{N_D}$$

$$R_{on} = \frac{4V_{BD}^2}{K_s \varepsilon_o \mu_n \varepsilon_{crit}^3}$$

where ε is the electric field, ε_{crit} the critical or breakdown electric field, V_{BD} the junction breakdown voltage, R_{on} the *on* resistance, W the space-charge region width or device thickness, and N_D the substrate doping concentration. For a uniformly-doped substrate the ε-x curve is linear with a slope proportional to the doping concentration N_D. The peak electric field in these figures is ε_{crit}. In Fig. 10(a), the ε-x curve is for a Si diode with N_{D1} and for a SiC diode with N_{D2}, where $N_{D2} > N_{D1}$. With the area under the curve for the SiC diode much larger than that of the Si diode, $V_{BD}(SiC) > V_{BD}(Si)$. Figure 10(b), a key figure, shows the case for $V_{BD}(SiC) = V_{BD}(Si)$, *i.e.*, for identical breakdown voltages, the SiC device is thinner with higher doping concentration, resulting in a much reduced *on* resistance. Experimental R_{on} data are shown in Fig. 11 versus breakdown voltage. Clearly R_{on} is much lower for SiC than for Si devices.

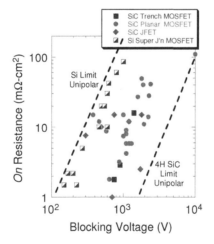

Fig. 11. *On* resistance versus junction breakdown for various semiconductor devices.[32]

3.1. *Schottky Diodes*

Standard Schottky diodes are shown schematically in Fig. 12(a). Schottky diodes typically have lower breakdown voltages than *pn* junction because of the space-charge region's sharp curvature at the device edge. The breakdown voltage is enhanced through field plates, floating guard rings, junction terminating extensions, and other methods. Providing a *p*-*n* junction at the metal edge, shown in Fig. 12(a), reduces this curvature leading to higher breakdown voltage. The *on* resistance can be reduced by introducing a mesh of p^+n junctions into the active area of the devices as shown in Fig. 12(b). Their primary function is to inject minority carriers in the case of a current surge causing conductivity modulation and lower R_{on}. In normal operation, *i.e.*, at nominal current ratings, the p^+n junctions are not active because the Schottky diode forward voltage drop is below the threshold of about 3 V for SiC p^+n junctions. Only in surge-mode operation do the p^+n structures inject, allowing for very high pulsed currents. This performance was achieved without affecting the good dynamic performance of conventional Schottky diodes.

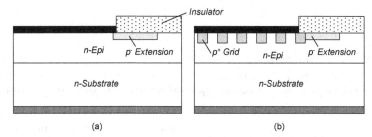

Fig. 12. Schottky diodes with (a) *p*-extension breakdown enhancer and (b) p^+ grid for *on* resistance reduction.

Diodes have a major impact on the overall efficiency of inverters and converters where they are mainly used as the anti-parallel freewheeling or snubber diodes to conduct the load current during insulated-gate bipolar transistor (IGBT) turn-off.[33] Such diodes eliminate the sudden voltage spike across an inductive load when its supply voltage is suddenly reduced or removed and the diode becomes forward biased. The snubber diode prevents this undesired voltage by conducting transient current around the device. For this application, a fast but soft diode reverse recovery is required. In addition to the voltage rating, current rating, and forward voltage drop, the recovery characteristics are important for fast switching power circuits and the recovery time and the waveform shape are affected by the diode and by the external circuit. For DC, a simple rectifier diode is often employed as another form of snubber. When current to the inductive load is rapidly interrupted, a large voltage spike is generated in the reverse direction. Placing the snubber diode in inverse parallel with the inductive load allows the current from the inductor to flow through the diode, dissipating the energy stored in the inductive load over the series resistance of the inductor and the usually much smaller resistance of the diode. Snubber diodes typically are not required for SiC MOSFETs.

Insulated-gate bipolar transistor (IGBT) turn-on losses increase as the reverse recovery rises and recovery time increases. To minimize the turn-on losses, the freewheeling diode must have fast and soft recovery. A diode with a snappy reverse recovery characteristic and hence high recovery dI/dt is problematic with transient voltages. Figure 13(a) compares the reverse recovery characteristics of a slow and snappy diode with a fast and soft recovery diode. The slow and snappy diode has a higher peak recovery current yet a higher dI/dt during the recovery phase. In contrast, the fast and soft diode has a reduced peak recovery current and dI/dt at the same time. Ideally, the freewheeling diodes of inverters/converters should have the following features[34]: low forward voltage and positive temperature coefficient for safe parallel diode operation, low reverse recovery losses, soft recovery, and ruggedness against dynamic avalanching, stable reverse blocking capability with low leakage current at high temperatures, and surge current capability, avalanche withstand capability, and a low forward recovery overshoot voltage during the diode turn-on transient period.

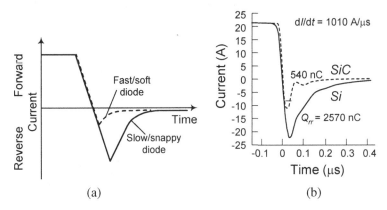

Fig. 13. (a) Diode reverse-recovery characteristics and (b) comparison of Si and SiC.

Schottky diode current is mainly carried by majority carriers with only a small minority carrier current component leading to the switching advantage that there is no need to remove excess carriers as must be done from the *n* region in *pn* diodes. Hence, there is a very low reverse-recovery current. Instead, only a displacement current for charging the metal-semiconductor junction capacitance of the diode is observed. On the other hand, with majority carrier current, there is very little conductivity modulation and the forward voltage drop or the *on* resistance tends to be high. The SiC Schottky current waveform depends only on the recharge speed of the junction capacitance with the charge transported by the displacement current being very low compared to the reverse recovery charge of *pin* diodes. Figure 13(b) shows the turn-off characteristics of three paralleled 4.5 kV SiC diodes and three paralleled 1.2 kV Si diodes.[35] In both measurements the same load current of 21 A was switched against a 600 V DC link voltage. The SiC diodes show much lower transient losses than state-of-the art 1.2 kV Si diodes. High voltage SiC diodes as freewheeling diodes in combination with silicon high-voltage IGBTs could save up to 50% of the switching losses in systems while conduction losses could be reduced even further. Reduced switching loss allows higher switching frequency operation, leading to smaller and less expensive passive components (inductors and capacitors). Schottky diodes are advertised for various ratings. One example is 1200 V, 60 A, 175°C.[36] Others lie in the 300-600 V range.[37]

3.2. *MOSFETs*

Since SiC MOSFETs will compete with Si MOSFETs, they need to be vertical devices to be able to handle the required currents and voltages. Lateral MOSFETs are unable to handle these high currents and voltages. Recently, enhancement-mode, vertical *n*-channel SiC MOSFETs have become commercially available. In these devices, in principle, there is no tail current during switching, resulting in faster operation and reduced switching loss. The low *on* resistance and compact chip size ensure low capacitance and low gate charge. These devices are rated at 1200 V with currents in the 20-30 A range and rated operating temperatures of 125°C.[38] These temperatures are well below the predicted temperatures of Fig. 1, however. Even though SiC is capable of high-temperature operation, the oxide, metallization and packages cannot handle such high temperatures. This, of course, limits their applications.

3.3. *Junction FETs*

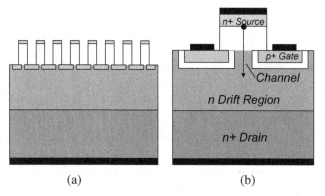

Fig. 14. (a) Cross section through a junction FET and (b) enlarged view of one cell.

In junction FETs (JFETs), shown in Fig. 14, the channel region is located away from the SiO$_2$/SiC interface leading to higher mobility than MOSFETs and low R_{on}. For example, the area-specific *on* resistance of a large-area 600 V unipolar JFET, is ~5 times lower than in charge-compensated silicon MOSFETs. The simplicity of the structure, which has only *pn* junctions as functional elements, ensures high ruggedness, *e.g.*, outstanding cosmic-ray resistance[39] and avalanche performance. With the channel electrons flowing through a bulk-like region, interface scattering, which reduces the mobility in MOSFETs, does not occur. Both normally-off and normally-on JFETs exist. In normally-off devices, drain current flows with zero gate voltage and the gate must be reverse biased to turn the channel off by the two space-charge regions in Fig. 14(b) touching each other. In normally-on devices, the channel is pinched off at zero gate voltage and must be opened by forward biasing the gate-substrate junction. This limits the drive voltage, but in SiC *pn* junctions forward current is very low until about 2.5-3 V when the diodes become forward biased. JFETs are advertised up to 1700 V, 4 A, and 175°C.[40]

4. The Competition

4.1. *Silicon*

The uses of power devices are illustrated in Fig. 15, clearly showing the applications in the lower frequency range to be dominated by Si. It is only at the higher frequencies and initially in the lower power domain that SiC plays a role. GaN shines at still higher frequencies, but lower powers.

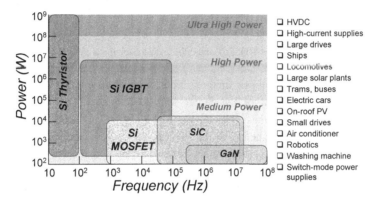

Fig. 15 Power vs. frequency for Si and wide energy gap devices.[41]

The main competitors of SiC and GaN devices are Si MOSFETs and IGBTs. *Silicon* MOSFETs became useful for power applications with the invention of the vertical MOSFET in Fig. 16(a). The channel length is determined by the lateral diffusion of the *p*-body and the n^+ source. The high current is made possible by the locating the drain at the bottom, rather than at the side, of the device. The high voltage is provided by decoupling the channel length from the n^+ drain. In spite of the high current capability of this device, the *on* resistance is limited by the fact that the device is a unipolar device with no conductivity modulation in the drift region. Reduction of the *on* resistance of power devices is extremely important. This is made possible by replacing the n^+ drain with a p^+ drain, in Fig. 16(b), in IGBTs, where holes are injected into the *n*-region from the p^+ drain leading to conductivity modulation.[42] This provides low *on* resistance with high current and high voltage. The downside is the presence of minority carriers in the *n*-region. During turn-off, these minority holes have to "disappear" reducing the frequency response. Various methods have been proposed and implemented. They usually involve some sort of minority carrier lifetime control.[43] Today's high-power IGBTs have voltage/current ratings ranging from 1700V/3600A to 6500V/750A and the main limiting factor for maximum output current is the free-wheeling diode in inverter and rectifier mode operation. A large portion of the power electronics market is satisfied with IGBTs. For example, the Toyota Prius hybrid automobile contains 18 IGBTs and 18 free-wheeling diodes. Further development requires diodes to match the IGBT and there is no need for improved switch generations unless the diode has reduced losses and higher power capabilities. That is where SiC diodes come in.

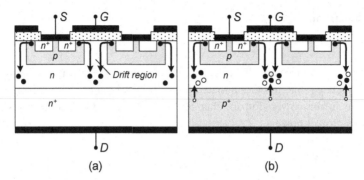

Fig. 16. (a) Power MOSFET and (b) insulated gate bipolar transistor.

As is frequently the case, improvements in basic devices come about due to demand. One of these is the invention of the Super MOSFET, also known as *COOLMOS*.[44] We explain this device with the simple diode in Fig. 17. Consider the p^+nn^+ diode in Fig. 17(a). This device is designed such that the reverse-biased space-charge region (scr) width, W_n, punches through the n-base and touches the n/n^+ boundary at breakdown. With constant base doping concentration, the ε-x curve is linear and the area under the curve represents the breakdown voltage V_{BD}. By reducing the n-region doping concentration, the slope of the ε-x plot is reduced and in the extreme *pin* diode, it has zero slope (Fig. 17(b)). Hence, for identical V_{BD}, the scr width is reduced from W_n to W_i leading to lower *on* resistance, which is extremely important for power devices.

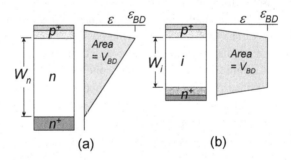

Fig. 17. (a) *pn* and (b) *pin* diodes device cross section and electric field plots.

The concept of Fig. 17 is implemented into vertical power MOSFETs in Fig. 18. Figure 18(a) shows a portion of a vertical MOSFET under reverse bias, *i.e.*, the drain is positively biased. The scr is contained in the region surrounding the p^+ body. By extending the body into the n-region in the super-junction MOSFET in Fig. 18(b), the scr volume is significantly increased resulting in higher V_{BD}. Alternatively, for a given V_{BD}, the n-region can be made thinner leading to reduced R_{on}. Such super-junction MOSFETs have the following properties: vertical *and* lateral space-charge regions $\Rightarrow$ reduced electric field, increased V_{BD} or reduced R_{on}, R_{on} reduced 5-10x, switch loss reduced 2x, N_D increased 10x for same V_{BD}, smaller conduction area $\Rightarrow$ higher current density.

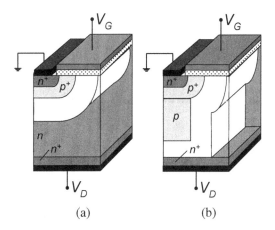

Fig. 18. (a) Conventional vertical MOSFET and (b) super-junction MOSFET.

4.2. *Gallium Nitride*

GaN devices have had initial success as RF switching devices, generally at lower voltages than those of SiC with GaN grown on sapphire substrates. The breakthrough for GaN came with the ability to grow GaN on (111)-oriented Si. These devices are usually high-electron mobility transistors (HEMT). GaN's higher switching speed and efficiency allow dc/dc converters to operate in the MHz region, saving space, reducing the need for heat sinks, and conserving power. Early GaN-on-Si devices had blocking voltages of 20 to 40 V for point-of-load dc/dc converters. These are depletion-type, normally-on FETs. More recent devices claim 200 V breakdown voltages with 600 V devices in the planning stage. One way to overcome the normally-on weakness is to pair a low-voltage Si MOSFET in a cascode configuration with a GaN HEMT, providing normally-off operation in a three-terminal device whose drive requirements are the same as a typical silicon-based power device.

5. Cosmic Ray Induced Failures

A new failure mode for high-current, high-voltage semiconductor devices was discovered in the early 1990s. It was of considerable practical significance and caused a series of equipment malfunctions. It consists of a sudden localized breakdown of a reverse-biased device with no detectable precursor.[45] The location of the breakdown spot on the wafer is random and such failures are generally lethal. The failure rate is constant in time but strongly depends on the applied voltage with a small dependence on temperature. Experiments in a German salt mine 140 m below ground did not show any of these failures, while experiments on the Jungfraujoch (3480 m above sea level) yielded a much higher failure rate than at sea level.[46] Irradiation with heavy energetic particles creates the same failure patterns. All together it was concluded that "cosmic rays" are the root cause of this kind of failure which is now supported by experiments done all around the world.

Primary cosmic rays are high-energy particles, mostly protons, in space that penetrate our atmosphere. They come from all directions and have a wide energy range and most originate from supernovae. A primary cosmic ray particle usually does not reach the earth's surface directly but collides with an atmospheric particle generating a variety of other particles, which later collide with other atmospheric particles. The process of a cosmic ray particle colliding with atmospheric particles and disintegrating into smaller pions, muons, neutrons, and the like, is called a cosmic-ray shower. Most of the generated particles are harmless for semiconductor devices but some, e.g., neutrons, are lethal.

One model suggests that the incident cosmic ray creates electron-hole pairs along its path, which perturb the electric field in the device space-charge region leading to local electric field enhancement and subsequent impact ionization. This, in turn, can lead to very high currents and filament formation and the resulting high power density may lead to thermal device breakdown.[47,48] Today's high-current, high-voltage power devices are designed for the failure rate due to cosmic rays to be reduced to an "acceptable" level. Nevertheless, such failures have to be taken into account for power electronic circuits. In particular, semiconductors for applications with a high utilization of the device's blocking capability and for equipment operating at high altitudes have to be assessed carefully. Such charged particles also are believed to be the cause of soft errors in integrated circuits. A soft error is a "glitch" in a semiconductor device, where the *ehps* generated by the interaction of an energetic charged particle with semiconductor atoms corrupts the stored information in the memory cell. These glitches are random, usually not catastrophic, and normally do not destroy the device.

Why are wide energy gap semiconductor devices less susceptible to cosmic ray damage? Electron-hole generation depends on the energy gap. Hence, the wider the energy gap, the lower is the *ehp* generation rate, making wide energy gap semiconductor devices more immune to cosmic radiation failure. The number of *ehps* generated, N_{ehp}, is[49]

$$N_{ehp} \approx \frac{E}{3.2E_G}\left(1 - \frac{\alpha E_{bs}}{E}\right)$$

where E is the energy of the incident particle/wave, E_G the energy gap, E_{bs} the mean energy of backscattered electrons and α the backscattering coefficient. This equations states that N_{ehp} is about 3x lower in SiC than in Si due to the wider energy gap.

6. Summary

Power semiconductor devices are largely made of silicon today, spanning Schottky diodes, pin diodes, MOSFETs, insulated-gate bipolar transistors, and thyristors. However, Si has progressed such that the material is very near its limit. Of the two most important power device parameters, on-resistance, R_{on}, and breakdown voltage, V_{BD}, Si can easily satisfy the V_{BD} requirements with breakdown voltages of 6000-8000 V available today. R_{on} is more difficult to reduce for Si but for wide energy gap semiconductor devices it is generally lower than that of Si, largely determined by the higher breakdown electric field leading to thinner substrates with its concomitant lower R_{on}. Furthermore, the typically lower carrier lifetimes of wide energy gap semiconductors leads to shorter turn-off times, higher frequency operation, and lower switching losses. The higher frequencies allow smaller passives (capacitors, inductors) and smaller volume systems. The higher thermal conductivity of SiC and its wider energy gap allow higher device temperatures and smaller heat sinks again leading to smaller volumes and reduced power consumption.

Wide energy gap semiconductors, SiC and GaN, have distinct advantages over Si, but they also have some problems. Although SiC has the ability to grow a thermal SiO_2, the SiO_2/SiC interface contains high interface trap densities and oxide charges leading to threshold voltage instabilities and contribute to the low carrier mobilities. SiC Schottky diodes, however, have been available commercially for over 10 years and are well received by the power system community. MOSFETs and junction FETs are being sampled.

GaN power device technology benefits from the extensive GaN material development for light-emitting devices, but it lacks native substrates. Although GaN FETs do not have the benefit of thermally-grown oxide, it does have heterojunction technology with its high electron mobility. Vertical device geometry, usually favored for power devices, is more difficult to implement, but lateral devices have been implemented. Breakdown voltages thus far are around 200 V, but will, no doubt, increase in the future.

In spite of the inherent material/device advantages, especially higher temperature and higher frequency capability, today's commercial devices are limited to ~200°C, largely governed by package considerations.

Acknowledgments

I thank those researchers who provided images and data.

References

1. H.C. Dunwoody, Wireless telegraph system, *US Patent* 837, 616 (1906).
2. H.J. Round, A note on carborundum, *Electrical World*, **19**, 309 (1907).
3. Charles & Colvard, Ltd, 300 Perimeter Park Drive Suite A, Morrisville, NC 27560, http://moissanite.com.
4. H. Moissan, Étude du siliciure de carbone de la météorite de Cañon Diablo, *Comptes Rendus Acad. Sci. Paris*, **140**, 405-406 (1905).
5. W. Shockley, *Proceedings of the First International Conference on Silicon Carbide,* Boston, MA, 1959 (Edited by J. R. O'Connor and J. Smiltens), Pergamon, New York (1960). I have edited this section slightly.
6. C. Buttay, D. Planson, B. Allard, D. Bergone, P. Bevilacqua, C. Joubert, M. Lazar, C. Martin, H. Morel, D. Tournier, and C. Raynaud, State of the art of high temperature power electronics, *Mat. Sci. Engin.* **B176**, 283-288 (2011).

7. P. Friedrichs, Silicon carbide power-device products – Status and upcoming challenges with a special attention to traditional, nonmilitary industrial applications, *Phys. Stat. Sol.* **B245**, 1232-1238 (2008).

8. R.K. Kirschman, in *High-Temperature Electronics*, IEEE Press, 3-35, 1999.

9. http://www.ssec.honeywell.com

10. S.P. Rountree, S. Berjaoui, A. Tamporello, B. Vincent, and T. Wiley, in: Proc. 5th Int. High Temp. Electronics Conf., Albuquerque, NM, (Sandia National Laboratories, Albuquerque, NM, 2000), p. I.P2.1 (2000).

11. P.G. Neudeck, S.L. Garverick, D.J. Spry, L.-Y. Chen, G.M. Beheim, M.J. Krasowski, and M. Mehregany, Extreme temperature 6H-SDiC JFET integrated circuit technology, *Phys. Stat. Sol.* **A206**, 2329-2345 (2009).

12. G. Majumdar and T. Oomori, Some key researches on SiC device technologies and their predicted advantages, *Power Electronics Europe*, Issue 6, 218-222 (2009).

13. R. Cheung, *Silicon Carbide Microelectromechanical Systems for Harsh Environments*, Imperial College Press, p. 3 (2006).

14. J.A. Lely, Darstellung von Einkristallen von Silicium Carbid und Beherrschung von Art und Menge der eingebauten Verunreinigungen, *Berichte der Deutschen Keramischen Gesellschaft* **32**, 229-236, (1955).

15. https://www.ifm.liu.se/semicond/new_page/research/sic/Chapter2.html.

16. S. Bai, R.P. Devaty, W.J. Choyke, U. Kaiser, G. Wagner and M.F. MacMillan, Determination of the electric field in 4H/3C/4H-SiC quantum wells due to spontaneous polarization in the 4H SiC matrix, *Appl. Phys. Lett.* **83**, 3171-3173 (2003).

17. P.K. Klein, Identification and carrier dynamics of the dominant lifetime limiting defect in n- 4H-SiC epitaxial layers, in *Silicon Carbide*, Vol. 1 (P. Friedrichs, T. Kimoto, L. Ley and G. Pensl, eds.), Wiley, Weinheim, 287-317 (2010).

18. K. Danno and T. Kimoto, Investigation of deep levels in *n*-type 4H-SiC epilayers irradiated with low-energy electrons, *J. Appl. Phys.* **100**, 113728-1 - 113728-6 (2006).

19. T. Kimoto, K. Danno and J. Suda, Lifetime-killing defects in 4H-SiC epilayers and lifetime control by low-energy electron irradiation, in *Silicon Carbide*, Vol. 1 (P. Friedrichs, T. Kimoto, L. Ley and G. Pensl, eds.), Wiley, Weinheim, 267-286 (2010).

20. G. Pensl et al., Alternative techniques to reduce interface traps in n- 4H-SiC MOS capacitors, in *Silicon Carbide*, 2 (P. Friedrichs, T. Kimoto, L. Ley and G. Pensl, eds.), Wiley, Weinheim, 193-214 (2010).

21. J.A. Cooper, Jr., Advances in SiC MOS technology, *Phys. Stat. Sol.* (a) **162**, 305-320 (1997).

22. H. Li, S. Dimitrijev, H.B. Harrison, and D. Sweatman, Interfacial characteristics of N_2O and NO nitrided SiO_2 grown on SiC by rapid thermal processing, *Appl. Phys. Lett.* 70, 2028-2030 (1997); G.Y. Chung, C.C. Tin, J.R. Williams, K. McDonald, M. Di Ventra, S.T. Pantelides, L.C. Feldman, and R.A. Weller, Effect of nitric oxide annealing on the interface trap densities near the band edges in the 4H polytype of silicon carbide, *Appl. Phys. Lett.* 76, 1713-1715 (2000); G.Y. Chung, C.C. Tin, J.R. Williams, K. McDonald, R.K. Chanana, R.A. Weller, S.T. Pantelides, L.C. Feldman, O.W. Holland, M.K. Das, and J.W. Palmour, Improved inversion channel mobility for 4H-SiC MOSFETs following high temperature anneals in nitric oxide, *IEEE Electron Dev. Lett.* **22**, 176-178 (2001); G.Y. Chung, J.R. Williams, T. Isaacs-Smith, F. Ren, K. McDonald, and L. C. Feldman, Nitrogen passivation of deposited oxides on *n* 4H-SiC, *Appl. Phys. Lett.* 81, 4266-4268 (2002); S. Dhar, Y.W. Song, L.C. Feldman, T. Isaacs-Smith, C.C. Tin, J.R. Williams, G. Chung, T. Nishimura, D. Starodub, T. Gustafson, and E. Garfunkel, Effect of nitric oxide annealing on the interface trap density near the conduction bandedge of 4H-SiC at the oxide/(1120) 4H-SiC interface, *Appl. Phys. Lett.* 84, 1498-1501 (2004).

23. S. Dhar, S. Wang, A.C. Ahyi, T. Isaacs-Smith, S.T. Pantelides, J. R. Williams and L.C. Feldman, *Mat. Sci. Forum,* **527-529**, 949 (2006).

24. D. Okamoto, H. Yano, K. Hirata, T. Hatayama and T. Fuyuki, Improved Inversion Channel Mobility in 4H-SiC MOSFETs on Si Face Utilizing Phosphorus-Doped Gate Oxide, *IEEE Electron Dev. Lett.* **31**, 710-712 (2010).

25. V.V. Afanas'ev and A. Stesmans, Hydrogen-induced valence alternation state at SiO_2 interfaces, *Phys. Rev. Lett.* **80**, 5176-5179 (1998); K. Fukuda, S. Suzuki, T. Tanaka, and K. Arai, Reduction of interface-state density in 4H-SiC *n*-type metal-oxide-semiconductor structures using high-temperature hydrogen annealing, *Appl. Phys. Lett.* **76**, 1585-1588 (2000); V.V. Afanas'ev, A. Stesmans, M. Bassler, G Pensl, and M. J. Schulz, "Comment on Reduction of interface-state density in 4H-SiC n-type metal-oxide-semiconductor structures using high-temperature hydrogen annealing [Appl. Phys. Lett. 76, 1585 (2000)]," *Appl. Phys. Lett.* **78**, 4043 (2001).

26. D. Kaplan, I. Solomon, and N.F. Mott, Explanation of the large spin-dependent recombination effect in semiconductors, *Journal De Physique Lettres*, **39**, L51-L54 (1978).

27. D.K. Schroder and J.A. Babcock, Negative bias temperature instability: A road to cross in deep submicron CMOS manufacturing, *J. Appl. Phys.* **94**, 1-18 (2003).

28. M. Bassler, V.V. Afanas'ev, G. Pensl, and M. Schultz, Degradation of 6H-SiC MOS capacitors operated at high temperatures, *Microelectron. Eng.* **48**, 257 (1999).

29. M.J. Marinella, D.K. Schroder, T. Isaacs-Smith, A.C. Ahyi, J.R. Williams, G.Y. Chung, J.W. Wan, and M.J. Loboda, Evidence of negative bias temperature instability in 4H-SiC metal oxide semiconductor capacitors, *Appl. Phys. Lett.* **90**, 253508-1-3 (2007).

30. J. Rozen, S. Dhar, S.T. Pantelides, L.C. Feldman, S. Wang, J.R. Williams and V.V. Afanas'ev, Suppression of interface state generation upon electron injection in nitrided oxides grown on 4H-SiC, *Appl. Phys. Lett.* **91**, 153503-1-3 (2007).

31. J. Rozen, S. Dhar, S.K. Dixit, V.V. Afanas'ev, F.O. Roberts, H.L. Dang, S. Wang, S.T. Pantiledes, J.R. Williams and L.C. Feldman, Increase in oxide hole trap density associated with nitrogen incorporation at the SiO_2/SiC interface, *J. Appl. Phys.* **103**, 124513-1-6 (2008).

32. T. Nakamura, M. Miura, N. Kawamoto, Y. Nakano, T. Otsuka, K. Okumura, and A. Kamisawa, Development of SiC diodes, power MOSFETs and intelligent power modules, *Phys. Stat. Solidi* **(a) 206**, 2403-2416 (2009).

33. Z.J. Shen and I. Omura, Power semiconductor devices for hybrid, electric and fuel cell vehicles, *Proc. IEEE*, **95**, 778-789 (2007).

34. V.K. Khanna, *IGBT Theory and Design*, Wiley-Interscience, NY, 60-62 (2003).

35. L. Lorenz and H. Mitlehner, Key power semiconductor device concepts for the next decade, *37th Industry Applic. Conf.* 564-569 (2002).

36. http://semisouth.com/wp-content/uploads/2011/05/DS_SDP60S120D_rev1.1.pdf.

37. http://search.digikey.com/scripts/DkSearch/dksus.dll?Cat=1376383&k=SiC.

38. http://www.cree.com/products/pdf/CMF20120D.pdf.

39. G. Soelkner, Reliability of SiC power devices against cosmic radiation-induced failure, European Conf. on SiC and Related Materials, Newcastle-upon-Tyne, Sept. 2006.

40. http://semisouth.com/power-semiconductors/sic-transistors.

41. Courtesy of R. Ploss, Infineon.

42. B.J. Baliga, M.S. Adler, P.V. Gray, R.P. Love and N. Zommer, The insulated gate rectifier (IGR): a new power switching device, *Tech. Digest IEEE IEDM* **82**, 264-267 (1982).

43. B.J. Baliga, Fast-switching insulated gate transistors, *IEEE Electron Dev. Lett.,* **EDL-4**, pp. 452-454 (1983).

44. G. Deboy, M. Marz, J.-P. Stengl, H. Strack, J. Tihanyi and H. Weber, A new generation of high voltage MOSFETs breaks the limit line of silicon, *Tech. Digest IEEE IEDM* **98**, 683-685 (1998).

45. H.R. Zeller, Cosmic ray induced failures in high power semiconductor devices, *Solid-State Electron.* **38**, 2041-2046 (1995).

46. N. Kaminski and T. Stiasny, Failure rates of IGCTs due to cosmic rays, Application Note 5SYA 2046-02, ABB Switzerland Ltd, 1-8 (2007).

47. H. Kabza, H.J. Schulze, Y. Gerstenmaier, P. Voss, J. Wilhelmi, W. Schmid, F. Pfirsch, and K. Platzoder, Cosmic radiation as a cause for power device failure and possible countermeasures, in *Proc. 6th Int. Symp. Power Semiconductor Devices IC's*, Davos, Switzerland, 9-12 (1994).

48. A.M. Albadri, R.D. Schrimpf, D.G. Walker, and S.V. Mahajan, Coupled electro-thermal simulations of single event burnout in power diodes, *IEEE Trans. Nucl. Sci.* **52**, 2194-2199 (2005).

49. J.F. Bresse, Quantitative investigation in semiconductor devices by electron beam induced current mode: a review, in *Scanning Electron Microscopy*, **1**, 717-725 (1978).

PERFORMANCE AND APPLICATIONS OF DEEP UV LED

M. SHATALOV, A. LUNEV, X. HU, O. BILENKO, I. GASKA, W. SUN, J. YANG,
A. DOBRINSKY, Y. BILENKO, R. GASKA and M. SHUR[1]

Sensor Electronic Technology Inc. 1195 Atlas Road, Columbia, South Carolina, 2920
shatalov@s-et.com
[1]*Center for Integrated Electronics, Rensselaer Polytechnic Institute, Troy, NY 12180, USA*

We discuss physics, design, fabrication, performance, and selected applications of Deep Ultraviolet Light Emitting Diodes (DUV LEDs). Our analysis reveals the relative contributions of electrical injection, internal quantum efficiency, and light extraction efficiency to the overall DUV LED performance. Our calculations show that the reduction of the dislocation density at least below value of 2×10^8 $1/cm^3$ is necessary for reaching high DUV LED efficiency. Better light extraction has been achieved using an innovative p-type transparent sub-contact layer and reflecting ohmic p-type contact resulting in nearly tripling DUV LED power. At high power dissipation, temperature rise might be significant, and we present data showing the power degradation with temperature increase and the results of the detailed 1D and 3D analysis of thermal impedance of DUV LEDs. As an example of DUV LED application, we report on microbial disinfection using 19 watt 275 nanometer DUV LED.

Keywords: Deep; Ultraviolet; Light Emitting Diode.

1. Introduction

Solid state ultraviolet light emitting diodes (UV LEDs) have numerous applications in biomedical industry. LEDs are used for purifying air and water, treating skin diseases, disinfecting surfaces, for industrial (photo-chemical) curing, printing, for forensic investigations and forgery detection [1–4]. Through the years, the UV electromagnetic radiation was primarily based on tube technology first developed in 1901 [5]. Though UV lamps are able to generate high output levels there are several drawbacks of UV lamps: lamps are fragile, can be a biological hazard, since they contain mercury and high voltages resulting in ozone production, require proper disposal techniques, have short lifetimes, and a fixed spectral power distribution. UV lamps cannot be dimmed. High pressure mercury lamps operate at high temperatures and may not be used in processes that require lower temperatures, for example, for photo-chemically curing polymers [6].

The organization of this paper is as following: in the first part of the paper we analyze the efficiency of deep UV LEDs (DUV LEDs) and compare their efficiency with visible LEDs. The details of DUV fabrication are outlined in the context of achieving optimal DUV LED efficiency. The thermal management is an important part of overall LED efficiency and is analyzed in subsequent section. Finally, we present the application of DUV LED for water disinfection and summarize the overall state of the art achievements in DUV LED in the concluding section.

2. DUV LED Efficiency

Light emitting diodes are very promising efficient light sources. Advanced visible LEDs can deliver more lumens per watt than any other lighting technology [7]. Both red [8] and blue [7] light emitting diodes demonstrated wall plug efficiencies of over 60%. The LED theory [9] and recent experiments [10] show that LED WPE could, in principle, exceed 100%.

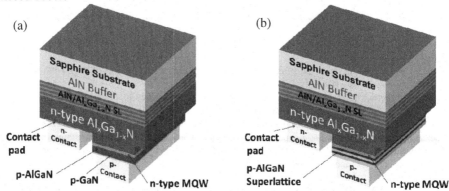

Fig. 1. LED schematics using flip chip design; (a) shows the typical DUV LED design, (b) illustrates advanced DUV LED design.

Figure 1 shows the general schematic of the LED flip chip design. The location of different regions is indicated on the figure. The Figure 1 (a) show a traditional device containing p-GaN layer and p contact with low reflectivity, whereas Figure 1 (b) shows an advanced device with p-AlGaN transparent layer and reflective p contact.

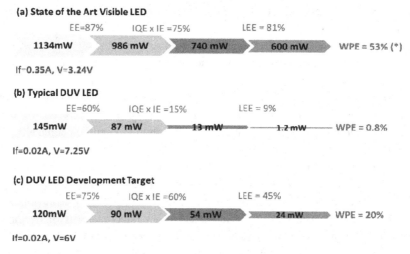

Fig. 2. Flow chart of losses during operation of visible and DUV LEDs. The overall wall plug efficiency (WPE) is indicated for each device. EE – electrical efficiency, IQE-internal quantum efficiency, IE-injection efficiency and LEE-light extraction efficiency; (a) – visible LED, (b) – typical DUV LED, (c) possible target DUV LED.

*Higher values for WPE has been reported for visible LED [7,8]

Detailed analysis of LED efficiency for visible LED is presented in Figure 2 (a) [11] and is compared to typical current wall plug efficiencies of DUV LED devices (Figure 2 (b)).

Typical DUV LEDs are considerably less efficient than their visible counterparts. DUV LED's decreased efficiency is due to poor transparency of semiconductor layers to UV light, poor UV light reflectivity of n- and p-contacts, and low conductivity of semiconductor heterostructure. Nevertheless, significant improvement has been achieved by fabricating more transparent semiconductor layers, by improving reflectivity of contacts, by chip encapsulation and by growth of semiconductor layers with low density of threading dislocations. All these improvements allow to predict possibility of achieving 20% wall plug efficiency for DUV LED as shown in Figure 2 (c).

The very basic and necessary improvement consists of increasing light extraction efficiency (LEE) of DUV LEDs.

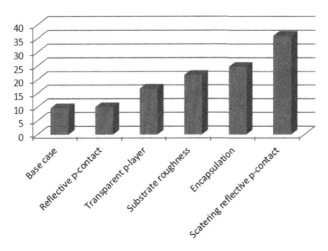

Fig. 3. Improvement of LEE due to transparency of semiconductor layers, improvement of reflectance of metal ohmic contacts, addition of surface roughness, chip encapsulation and contact scattering.

The increased LEE is demonstrated in Figure 3, where effects of transparency of semiconductor layers, reflectivity of contacts, introduction of surface roughness, encapsulation and scattering of contacts are evaluated using ray tracing. It has been already seen that some of these improvements yield great results, with current state of art DUV LED's LEE's being on the order of 30%.

Transparency of semiconductor layers is achieved by employing superlattices of $Al_xGa_{1-x}N/Al_yGa_{1-y}N$ type with sufficiently high molar fraction of aluminum to reduce semiconductor layer absorbance. These layers are schematically illustrated in Figure 1(b) and are labeled as p-AlGaN layers. The delicate balance is needed to maintain high mobility and high carrier concentration density in p-AlGaN layers and at the same time maintain low absorbance. Figure 4 demonstrates the transmission properties of the semiconductor superlattices. Without presence of p-AlGaN superlattice layers the absorbance of p-contact is 90% or higher for DUV LEDs operating at wavelengths less than 300nm. p-AlGaN superlattices reduce absorption to as low as 30%.

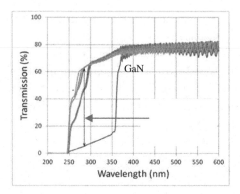

Fig. 4. Transmission of the superlattices with various content of aluminum in quantum wells and barriers.

In Figure 4, we include the normal incidence optical transmission spectra of LED structures with aluminum composition in p-AlGaN layer varying from 0 to 35%. As follows from optical modeling of the LED structure, increase of the transmission from 5% to approximately 60% corresponds to the reduction of the optical losses in the p-layer from 10^5 cm^{-1} to below 1000 cm^{-1}. A UV reflective p-metal along with reduced optical loss enables multiple pass light extraction and resulted in a factor of 2 increase of output power from LED die.

While Figure 3 demonstrates improvements in LEE, some of the main challenges in LED light extraction are related to the light trapping inside semiconductor layers due to total internal reflection. Simulations done with ray tracing confirm light trapping in the inner semiconductor layers. Significant trapping typically happens at the layer interfaces with large discontinuity in refractive index and for cases when index of refraction is decreased as light traverses an interface. A large discontinuity in the refractive index is observed at the interface of AlN buffer layer and sapphire wafer, for example, where the interface allows roughly only half of all the light to be transmitted, while the rest of the light flux is reflected back towards absorbing semiconductor layers and contacts. Also, interfaces between AlGaN and AlN semiconductor layers produce additional light trapping that accounts for a substantial subsequent loss of the emitted light. Various roughness technics have been proposed [12–14] and currently are investigated at SET, Inc. to partially remedy this problem.

Increasing internal quantum efficiency (IQE) provides another avenue for DUV LED improvement. To increase IQE, high quality semiconductor layers must be fabricated with low levels of threading dislocations and point defects. Figure 5 illustrates the effect of threading dislocations and point defects on the IQE of a typical group III nitride LED. The theoretical calculations of Figure 5 are based on the model of Karpov et al. [15] and confirmed by experiments [16]. As can be seen from Figure 5, low dislocation densities and point defects (large values of τ_{NR}) are required in order to obtain high IQE's.

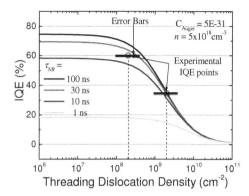

Fig. 5. Effect of threading dislocations and point defect on IQE of the LED device.

3. DUV LED Fabrication

To achieve high transparency superlattices and low dislocation density semiconductor layers innovative techniques were employed to grow semiconductor layers in the DUV LED structure.

Epitaxial structures were grown on c-plane sapphire substrates by a combination of conventional metal-organic chemical vapor deposition (MOCVD) and migration enhanced metal-organic chemical vapor deposition (MEMOCVD®) [17]. To reduce TDD in the active region, we used 1 μm thick high-quality AlN layer grown by MEMOCVD®. Aluminum composition was graded down using strain-relief AlN/AlGaN superlattice to allow for the deposition of a Si-doped bottom AlGaN cladding/contact layer. This was followed by growth of the MQW structure capped with the Mg-doped p-AlGaN top cladding/contact layer. Cross-section and plane-view transmission electron microscopy confirmed TDD in MQW region to be below 2×10^8 cm^{-2} for all types of dislocations [18].

After growth, the wafers were processed with 350 μm x 350 μm LED devices optimized for uniform current spreading [19]. First, mesa geometries were defined with reactive ion etching. This was followed with the deposition of the n-contact layer based on the Cr/Ti/Al metal stack [20] annealed above 850 °C in N$_2$ ambient. For n-Al$_{0.6}$Ga$_{0.4}$N layers used in the 278 nm emission structures, measurements of linear transfer length model (TLM) pattern yielded the specific contact resistance in the range of $2\text{-}5 \times 10^{-5}$ Ω·cm^2. After the n-contact annealing, the p-contact based on reflective metal stack was deposited and annealed in air ambient at 500 °C [21]. This was followed by deposition of the probe contact metal, the dielectric for surface passivation and thick bonding pad metal. After processing, DUV LED chips were cingulated and mounted on TO-39 packages.

Reflecting contacts were composed of several metallic layers for reaching a balance between good conductivity of ohmic contact and the good reflectivity of the metal contact assembly. Figure 6 shows the improvement in reflectivity with the new reflective contact design based on multiple metallic layers. In this contact structure the base layer provides good adhesion, the top layer is highly reflective and the middle layer decreases contact resistance. Optimization of this contact structure should be further improved in order to reach the target efficiency shown in Figure 2.

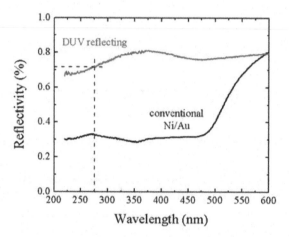

Fig. 6. Reflectivity of metal contacts as a function of wavelength.

The reflectivity of about 70% was achieved in UV spectra for novel design. The effect of transparent contacts is further illustrated by Figure 7, where the output power is plotted as a function of an applied current. Figure 8 shows the extraction quantum efficiency EQE and wall plug efficiency (WPE).

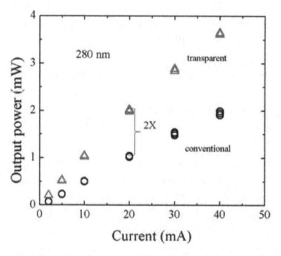

Fig. 7. Power output as a function of applied current. The red symbols correspond to LED with transparent contact.

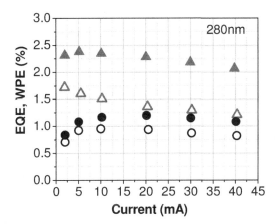

Fig. 8. External quantum efficiency and wall plug efficiency as a function of applied current. EQE is given by solid symbols, while WPE is shown by open symbols. The red symbols correspond to LED with transparent contact.

4. Thermal Analysis of DUV LED

DUV LED are still operating at a relatively low efficiency, thus large amount of LED driving power is dissipating in heat. This makes thermal management a key issue for DUV LEDs. Figure 9 demonstrates how the performance of DUV LED is affected by high junction temperatures.

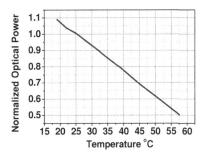

Fig. 9. Effect of junction temperature on device optical power. The optical power is normalized at the room temperature [28].

In this section, we present a steady state thermal analysis of LED device by first investigating the thermal resistance of LED chip layers and interfaces, and then considering the thermal resistance of the components of the LED package. According to a one dimensional model, the thermal resistance is given by

$$R_t = L/(\kappa A) \tag{1}$$

where L is the thickness of a layer, A is the cross sectional area and κ is the thermal conductivity of the layer. Figure 10 shows different semiconductor layers of typical LED die, while Figure 11 shows a typical LED package.

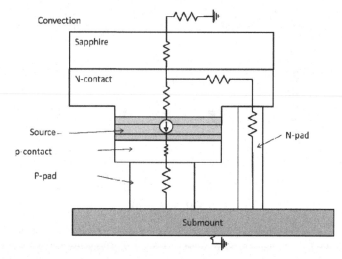

Fig. 10. Layers and thermal resistance elements of a typical LED die.

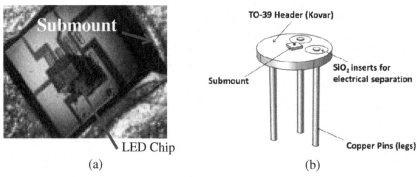

Fig. 11. Typical LED package [28].

Using 1-D model the thermal resistance of LED semiconductor layers, interfaces, contacts, submount and package are calculated and listed in the 1-D diagram in Figure 12.

Based on Figure 12, the total thermal resistance of the device is given by the following:

$$R_{TH} = R_1 + \cfrac{1}{\cfrac{1}{R_2} + \cfrac{1}{R_3}} + R_4 \tag{2}$$

where $R_1 \sim 3.5 \cdot 10^5 \left[\dfrac{K}{W}\right]$, $R_2 \sim 3.8 \left[\dfrac{K}{W}\right]$, $R_3 \sim 1000 \left[\dfrac{K}{W}\right]$, $R_4 \sim 20.3 \left[\dfrac{K}{W}\right]$, $R_{TH} \sim 24 \left[\dfrac{K}{W}\right]$.

While some layers can be well approximated using one-dimensional models others are inherently three dimensional and require corresponding three-dimensional heat transfer

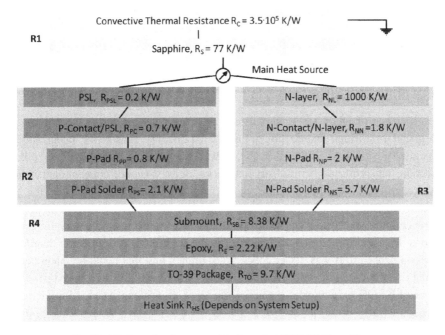

Fig. 12. 1-D thermal resistance model of a typical DUV LED based lamp.

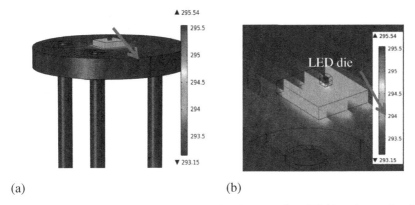

(a)   (b)

Fig. 13. 3-D simulation of temperature distribution of LED mounted on TO-39 package using COMSOL Multiphysics, (a) temperature distribution on the package surface, (b) temperature field slices through the device.

modeling. In this work we have utilized COMSOL Multiphysics to model joule heating of the device. Figure 13 illustrate heat distribution of an LED with TO-39 package.

Both the submount and TO-39 have to be modeled taking into account three dimensional nature of heat transfer in the device. The effective thermal resistances for submount and TO-39 in Figure 12 were calculated based on 3-dimensional COMSOL modeling.

In COMSOL modeling, the detailed heat transfer inside LED die was ignored and the LED die was modeled as an electrical resistor with the thermal resistance estimated using the 1D model. The resistance and applied voltage of the LED die were selected to match the typical operational powers dissipated by the LED device.

5. DUV LED Sterilization

While efficiency of DUV LED can still be improved, they are already widely used in industry and in various disinfection units [22], [23]. The reason for using LED for disinfection is manifold. First, the DUV LED sterilization is superior to chlorine disinfection that produces harmful organic by-products. Second, conventional mercury lamps are brittle, bulky, and contain toxic chemicals, and are not suitable for harsh environments. Third, DUV LEDs are best suited for developing countries where there is a scarce access to clean water and water purifications systems are necessary to suppress infectious diseases. SETi designed and fabricated multiple DUV LED water disinfection chambers. Intense effort was placed toward designing a unit that could be integrated into everyday Class B drinking water systems. Testing was geared toward achieving the NSF-55 standard 6 LOG reductions of bacterial microbes in drinking water. Multiple tests at various flow rates were performed. The test bacteria were *Escherichia Coli* and *Enterococcus*. Figure 14 demonstrates the use of DUV LED lamp for disinfection as it applies to reducing levels of E.coli and Enterococcus. The figure shows logarithmic reduction of *Enterococcus* and *E.coli* with a 19W lamp as a function of water flow. Figure 14 shows an almost 7 LOG reduction of *E.coli* at 0.5 liters per minute and a 5 LOG reduction of *Enterococcus* at 19W. As expected, the disinfection efficacy of the

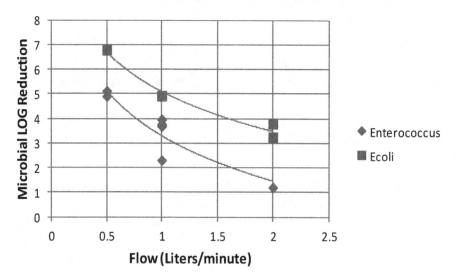

Fig. 14. LOG reduction of Enterococcus and E.coli with at 19W with respect to flow.

UV Adaptor and Carbon Block E.coli disinfection

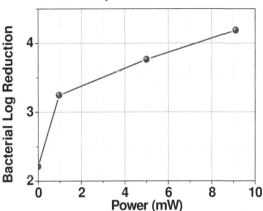

Fig. 15. Use of UV adaptor and carbon block filter for E.coli disinfection.

chamber declined with increased flow rate up to 2 liters per minute. SETi also looked into expanding the use of DUV LED's into already existing water disinfection systems. Standard carbon block filters were tested and shown to be ineffective in preventing high bacterial concentrations on their own. The incorporation of DUV LED's into a carbon block and UV disinfection system increased the disinfection rate by at least 99%. Figure 15 illustrates reduction *of E.coli* when a single UV lamp and carbon block filter were used together.

6. Conclusion

In summary, significant progress has been made to improve efficiency of DUV LED devices. New methods of growth of semiconductor layers lead to reduction in threading dislocation density and result in IQE increase. Further improvements in optical light extraction lead to the overall increase of extraction efficiency of LED devices. Thermal analysis helps identify high thermal resistivity regions in the device and is especially important for design of efficient high power DUV LED lamps. Finally we show that DUV LEDs are efficient light sources for sterilization.

Acknowledgments

Work at Sensor Electronic Technology was partially supported by the Defense Advanced Research Program Agency (Dr. J. Albrecht) through cooperative research agreement W911NF-10-2-0023 with the U.S. Army Research Laboratory (Dr. M. Wraback). The Authors also would like to thank Dr. Z. Liliental-Weber from Lawrence Berkley National Laboratory for TEM analysis of epitaxial structures. The work at RPI was partially supported by the subcontract from SET, Inc. and by the National Science Foundation under Cooperative Agreement EEC-0812056 and by the New York State Foundation for Science, Technology and Innovation (NYSTAR) under contracts C080145 and C090145.

Appendix A.   Calculation of Thermal Resistances

Thermal resistances for conductive semiconductor layers are defined by: $\Delta Q = \Delta T/R$, with R-thermal resistance being $R = \Delta L/(\kappa \cdot A)$, with ΔL – being thickness of semi-conductor layer, κ – thermal conductivity of a layer and A – area of the layer. Figure A1 shows the typical geometry of the device, and Table A1 lists the parameters of the device used to obtain thermal resistance of layers and interfaces. These thermal resistances are listed in Figure 12.

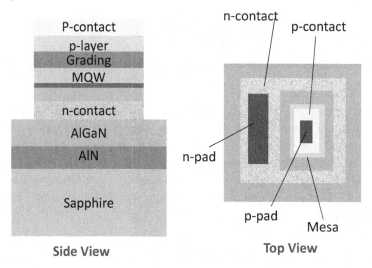

Fig. A1. Shows typical device geometry.

Table A1. Thermal conductivity and thermal resistance of various layers and components used in the model.

Element	Thermal Conductivity	Thermal Resistance
	[W/(m·K)]	[K/W]
Sapphire	3.50E+01	7.70E+01
PSL	3.00E+01	2.02E-01
P-Pad	4.00E+02	3.83E-01
P-Solder	5.70E+01	2.15E+00
P-Contact/PSL (approx.)	4.00E+02	6.90E-01
N-Contact/N-layer (approx.)	4.00E+02	1.84E+00
N-Solder	5.70E+01	5.74E+00
N-Pad	4.00E+02	1.02E+00
N-layer vertical	3.00E+01	2.99E-01
N-layer		1.14E+03
Submount 3D		8.38E+00
Epoxy	1.00E+01	2.22E+00
TO-39 3D		9.70E+00

Appendix B.   Thermal Conductivity of Al$_x$Ga$_{1-x}$N Semiconductor

Thermal conductivity of Al$_x$Ga$_{1-x}$N as a function of the temperature of the device was fitted based on the available experimental data [24], [25]. Figure B1 shows experimental data for AlN and GaN semiconductor materials.

The thermal conductivity fit for GaN (Figure B1 (a)) is approximated by the following expression $\kappa_{GaN}(\tau) = 4300e^{-10.5\tau} + 180e^{-0.39\tau}$, $(0.3 < \tau < 1.3)$, and for AlN (Figure B1 (b))

$$\kappa_{AlN}(\tau) = \frac{-7.2\tau^2 + 150\tau - 0.03}{\tau^2 - 0.5\tau + 0.3}, \ (0.1 < \tau < 3.5), \ \tau = \frac{T}{T_0}, \ T_0 = 300\text{K},$$

with all the coefficients having units of W/(m·K) when needed.

In order to fit thermal conductivity for Al$_x$Ga$_{1-x}$N we use the fit given by [26] page 23 and by [27],

$$\kappa_{Al_xGa_{1-x}N}(x,\tau) = \left[\frac{x}{\kappa_{AlN}(\tau)} + \frac{1-x}{\kappa_{GaN}(\tau)} + \frac{x(1-x)}{C_{AlGaN}} \right]^{-1} \tag{B.1}$$

Figure B2, shows the behavior of thermal conductivity as a function of aluminum mole fraction x.

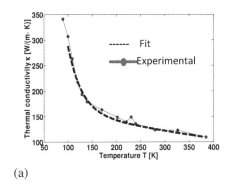

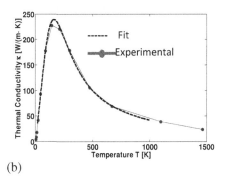

(a)　　　　　　　　　　　　　　　　(b)

Fig. B1. Fit to experimental data using conservative thermal conductivity values for different values of the temperature; (a) – fit for GaN [24], (b) – fit for AlN [25].

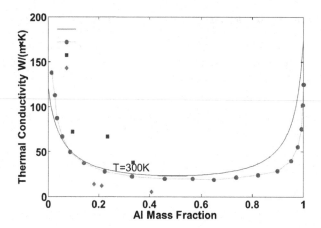

Fig. B2. Fit to experimental data. Solid black curve fit using equation A.1, green curve with solid circles – theoretical results [29], blue rectangle – experimental results [29], red diamonds – experimental results [30].

References

1. M. S. Shur and R. Gaska, "Deep-Ultraviolet Light-Emitting Diodes," *IEEE Transactions on Electron Devices*, vol. 57, no. 1, pp. 12–25, 2010.
2. D. Birtalan, "Compound Semiconductor," vol. 16, no. 2, pp. 28–30, Feb. 2010.
3. D. Birtalan, *Optoelectronics: infrared-visible-ultraviolet devices and applications*, 2nd ed. Boca Raton: CRC Press, 2009.
4. M. Shur, M. Shatalov, A. Dobrinsky, and R. Gaska, "Deep Ultraviolet Light-Emitting Diodes," in *GaN and ZnO-based Materials and Devices*, vol. 156, S. Pearton, Ed. Berlin, Heidelberg: Springer Berlin Heidelberg, 2012, pp. 83–120.
5. P. Hewitt, "Method of Manufacturing Electric Lamps," *US Patent. No. 682692*, Issue date: Sep 17, 1901.
6. S. Volk, "Phoseon Technology Ships the Highest Performing UV LED Curing Lamps on the Market," http://www.phoseon.com, March 7 2012, Phoseon Technology http://www.phoseon.com/Press%20Releases/FirePower_Shipping_Now_March2012.pdf
7. G. Meneghesso, "Is reliability of LEDs sufficient for massive market penetration?" *Journal of Physics D: Applied Physics*, http://iopscience.iop.org/0022-3727/page/Recent%20results%20on%20the%20degradation%20of%20white%20LEDs%20for%20lighting
8. News, Osram, "Osram red LED prototype breaks 200lm/W efficiency barrier" *Semiconductor Today*, 11 October 2011, http://www.semiconductor-today.com/news_items/2011/OCT/OSRAM_111011.html
9. A. Zukauskas, M. Shur, and R. Gaska, *Introduction to solid-state lighting*. New York: Wiley-Interscience, 2002, ISBN: 978-0-471-21574-5.
10. P. Santhanam, D. Gray, and R. Ram, "Thermoelectrically Pumped Light-Emitting Diodes Operating above Unity Efficiency," *Physical Review Letters*, vol. 108, no. 9, p. 097403, Feb. 2012.
11. A. Laubsch, M. Sabathil, J. Baur, M. Peter, and B. Hahn, "High-Power and High-Efficiency InGaN-Based Light Emitters," *IEEE Trans. Electron Devices*, vol. 57, no. 1, pp. 79–87, Jan. 2010.

12. Y.-C. Lee, C.-H. Ni, and C.-Y. Chen, "Enhancing light extraction mechanisms of GaN-based light-emitting diodes through the integration of imprinting microstructures, patterned sapphire substrates, and surface roughness," *Opt Express*, vol. 18, Suppl. 4, pp. A489–498, Nov. 2010.

13. C. B. Soh, B. Wang, S. J. Chua, V. K. X. Lin, R. J. N. Tan, and S. Tripathy, "Fabrication of a nano-cone array on a p-GaN surface for enhanced light extraction efficiency from GaN-based tunable wavelength LEDs," *Nanotechnology*, vol. 19, no. 40, p. 405303, Oct. 2008.

14. J.-Y. Kim, M.-K. Kwon, K.-S. Lee, S.-J. Park, S. H. Kim, and K.-D. Lee, "Enhanced light extraction from GaN-based green light-emitting diode with photonic crystal," *Appl. Phys. Lett.*, vol. 91, no. 18, p. 181109, 2007.

15. S. Y. Karpov and N. Makarov, "Dislocation effect on light emission efficiency in gallium nitride," *Applied Physics Letters*, vol. 81, p. 4721, Dec. 2002.

16. K. Ban, J. Yamamoto, K. Takeda, K. Ide, M. Iwaya, T. Takeuchi, S. Kamiyama, I. Akasaki, and H. Amano, "Internal Quantum Efficiency of Whole-Composition-Range AlGaN Multiquantum Wells," *Applied Physics Express*, vol. 4, p. 2101, May 2011.

17. R. Gaska, J. Zhang, and M. Shur, "Layer growth using metal film and/or islands," U.S. Patent US 7491626, 17-Feb-2009.

18. Z. Liliental-Weber: private communication.

19. Y. Bilenko, R. Gaska, and M. Shur, "Shaped contact layer for light emitting heterostructure," U.S. Patent 7928451, 19-Apr-2011.

20. R. Gaska, X. Hu, and M. Shur, "Chromium/titanium/aluminum-based semiconductor device contact," U.S. Patent Application US 2008/0315419 A1, 25-Dec-2008.

21. R. Gaska, M. S. Shatalov, and M. Shur, "Deep ultraviolet light emitting diode," U.S. Patent US 2011/0309326 A1, 22-Dec-2011.

22. S. Vilhunen, H. Särkkä, and M. Sillanpää, "Ultraviolet light-emitting diodes in water disinfection," *Environmental Science and Pollution Research*, vol. 16, no. 4, pp. 439–442, Feb. 2009.

23. M. A. Würtele, T. Kolbe, M. Lipsz, A. Külberg, M. Weyers, M. Kneissl, and M. Jekel, "Application of GaN-based ultraviolet-C light emitting diodes – UV LEDs – for water disinfection," *Water Research*, vol. 45, no. 3, pp. 1481–1489, Jan. 2011.

24. W. Liu, A. A. Balandin, C. Lee, and H.-Y. Lee, "Increased thermal conductivity of free-standing low-dislocation-density GaN films," *physica status solidi (a)*, vol. 202, p. R135–R137, Sep. 2005.

25. K. Watari, H. Nakano, K. Urabe, K. Ishizaki, S. Cao, and K. Mori, "Thermal conductivity of AlN ceramic with a very low amount of grain boundary phase at 4 to 1000 K," *Journal of Materials Research*, vol. 17, no. 11, pp. 2940–2944, Jan. 2011.

26. R. Quay, *Gallium nitride electronics*. Berlin: Springer, 2008, ISBN 9783540718901

27. V. Palankovski and S. Selberherr, "Thermal models for semiconductor device simulation," pp. 25–28, HITEN 99. Third European Conference on High Temperature Electronics. (IEEE Cat. No.99EX372) 7 July 1999.

28. Sensor Electronic Technology, *UVTOP, Deep UV LED Technical Catalogue*, http://www.s-et.com/uvtop-catalogue.pdf

29. W. Liu and A. A. Balandin, "Thermal conduction in $Al_xGa_{1-x}N$ alloys and thin films," *Journal of Applied Physics*, vol. 97, no. 7, p. 073710, 2005.

30. B. C. Daly, H. J. Maris, A. V. Nurmikko, M. Kuball, and J. Han, "Optical pump-and-probe measurement of the thermal conductivity of nitride thin films," *Journal of Applied Physics*, vol. 92, no. 7, p. 3820, 2002.

ORDERED GaN/InGaN NANORODS ARRAYS GROWN BY MOLECULAR BEAM EPITAXY FOR PHOSPHOR-FREE WHITE LIGHT EMISSION

S. ALBERT[*], A. BENGOECHEA-ENCABO[*], M.A. SANCHEZ-GARCÍA,
F. BARBAGINI and E. CALLEJA[†]

ISOM-Dept. Ing. Electrónica, ETSIT, Univ. Politécnica, 28040 Madrid, Spain
[†]calleja@die.upm.es

F. LUNA, A. TRAMPERT and U. JAHN

Paul-Drude-Institut, Hausvogteiplatz 5-7, 10117 Berlin, Germany

P. LEFEBVRE

ISOM-Dept. Ing. Electrónica, ETSIT, Univ. Politécnica, 28040 Madrid, Spain
and
Groupe d'Etude des Semiconducteurs, Univ. Montpellier II, 34095 Montpellier, France

L.L. LÓPEZ, S. ESTRADÉ, J.M. REBLED and F. PEIRÓ

LENS, MIND-IN2UB, Dept. Electrónica, Univ. de Barcelona, 08028 Barcelona, Spain.

G. NATAF, P. DE MIERRY and J. ZUÑIGA-PÉREZ

CRHEA-CNRS, Rue Bernard Gregory, 06560 Valbonne, France

The basics of the self-assembled growth of GaN nanorods on Si(111) are reviewed. Morphology differences and optical properties are compared to those of GaN layers grown directly on Si(111). The effects of the growth temperature on the In incorporation in self-assembled InGaN nanorods grown on Si(111) is described. In addition, the inclusion of InGaN quantum disk structures into self-assembled GaN nanorods show clear confinement effects as a function of the quantum disk thickness. In order to overcome the properties dispersion and the intrinsic inhomogeneous nature of the self-assembled growth, the selective area growth of GaN nanorods on both, c-plane and a-plane GaN on sapphire templates, is addressed, with special emphasis on optical quality and morphology differences. The analysis of the optical emission from a single InGaN quantum disk is shown for both polar and non-polar nanorod orientations.

Keywords: Nanorods; GaN; InGaN; SAG; white light.

1. Introduction

Group III-nitride semiconductors are very appealing for applications in optoelectronic devices, in particular, InGaN alloys for light emitters working in the whole visible spectral region,[1,2,3] that open the way to develop phosphor-free white light emitters.

Most research efforts focused mainly on c-plane quantum well (QW) structures, though their efficiency suffers from (i) a reduced radiative recombination rate due to the quantum confined Stark effect derived from spontaneous and piezoelectric polarizations,

---

[*] These authors contributed equally to this work.

and (ii) the high density of nonradiative defects due to the lattice mismatch between GaN and InGaN.

It has been shown that dislocation- and strain- free group-III nitrides can be grown on Si(100) and (111), as well as on amorphous SiO$_2$ substrates[4,5,6,7,8] in the form of one-dimensional structures such as nanorods (NRs). A potential benefit of NRs as light emitters, aside from their very high crystal quality, relies on better light extraction efficiency as compared to thin films, because of a guiding effect along the NR axis. When embedding quantum disks (QDisc) in NRs, light emission enhancement is expected due to carrier confinement.[9,10,11,12,13] In addition, it has been demonstrated that white light emission can be obtained from self-assembled InGaN/GaN NRs.[14,15,16,17]

Self assembled arrays of InGaN/GaN nanocolumnar Light Emitting Diodes (LEDs) always show polychromatic emission mainly due to an inhomogeneous distribution of NRs geometry (diameter, height), In content, and strain distribution within the active region (QDisc).[15,16,18] In addition, self-assembled NRs may merge generating extended defects as dislocations and stacking faults. Thus, efficient white LEDs based on III-nitrides grown by a self-assembled method have little chances to become commercial. However, self-assembled III-nitride NRs played a significant role in helping to understand material properties and the physics involved in efficient light emission from almost defect-free nanocrystals.

The first part of this work will address the growth of self-assembled GaN NRs on Si(111) substrates, as well as InGaN/GaN NRs and InGaN QDiscs embedded into GaN NRs. The second part of the work will address the selective area growth (SAG) of these structures leading to ordered arrays as functional blocs to develop actual devices (LEDs). During the last years the SAG of GaN NRs[19,20,21,22] has been studied by a number of groups, allowing the fabrication of NRs with well controlled position and diameter. In this work, SAG of GaN NRs on polar (c-plane) and non-polar (a-plane) sapphire/GaN templates will be addressed. The incorporation and properties of InGaN QDiscs to both polar and non-polar NRs will be studied.

2. Experimental

The plasma assisted molecular beam epitaxy (PAMBE) system used in this work was equipped with a rf-plasma source providing active nitrogen and standard Knudsen cells for Ga and In. All samples were grown, either on Si(111) for self-assembled GaN NRs, or on titanium nanohole masks deposited on c-plane or a-plane GaN/sapphire templates for SAG of GaN NRs. Metal and nitrogen fluxes were calibrated in (0001) GaN and (0001) InN growth rate units (nm/min).[23] In wurtzite GaN and InN the areal densities referring to 1 monolayer (ML) are 1.14x10^{15} GaN/cm^2 and 9.17x10^{14} InN/cm^2 respectively.[24,25] Once grown, the samples were studied with scanning electron microscopy (SEM), photoluminescence (PL), and transmission electron microscopy (TEM). PL experiments were carried out using a He-Cd laser with a constant power of 30 mW as excitation source. Due to the chromatic collection system used, different signal optimizations in the case of a broad spectrum are possible, thus potentially deforming the overall aspect of the spectrum. Then, the collected optical signal was systematically optimized by setting the spectrometer at the center of the visible spectrum, therefore slightly detrimental to contributions from the red and blue-violet regions.

3. Results and Discussion

3.1. *Growth of self-assembled nanocolumns on Si(111) substrates*

In the last years many reports on the growth of self-assembled III-nitride NRs have been published.[26,27,28,29,30,31] The basic principles that govern the self-assembled growth process may now be summarized in a brief way yielding the necessary information to understand them. This will allow later a straight comparison with SAG.

Self-assembled GaN NR grown on Si(111) have excellent structural and optical properties, as compared to those of thin films. The reason for that relies on the absence of extended defects in the NRs that are also intrinsically free of strain because of their small footprint and large surface to volume ratio. This allows NRs to "disconnect" from the substrate after a few MLs by generating dislocations at the interface.[32] The self-assembled growth of GaN NRs is governed by two main factors: (i) the growth temperature and (ii) the Ga/N ratio. The impact of these parameters on the growth has been studied extensively.[24]

The actual Ga/N ratio for the growth of self-assembled NRs has to be smaller than 1 in order to prevent the coalescence (enhanced lateral growth) of the isolated GaN islands formed at the very beginning of the growth. Note that the actual Ga/N ratio may substantially differ from the nominal one, because of Ga desorption and GaN decomposition that may have a strong effect depending on the temperatures used.[23,33]

Assuming constant Ga and N fluxes of Φ_{Ga} = 1 nm/min and Φ_N = 6 nm/min that yield a III/V ratio smaller than 1, lower and upper values of the growth temperature allowing the NRs growth (NR regime) can be defined. The lower temperature limit is determined by the transition from self-assembled NR regime to compact layers, at around 760°C (for the established fluxes), that can be understood in terms of Ga diffusion which must be high enough in order to allow for the nucleation of well separated GaN islands.[30] The upper temperature limit is determined by the balance between the nominal growth rate and the Ga desorption and GaN decomposition rates, both being exponential functions of the temperature. Beyond this limit no growth is possible.[33] Note that this boundary, that occurs around 810°C for the established fluxes, shifts towards higher temperatures when increasing the Ga flux.

Figure 1 shows SEM images of GaN samples grown directly on Si(111) at different temperatures and Ga/N ratios. Figure 1a refers to a GaN layer of around 700 nm thickness grown at 750°C with a Ga-flux (Φ_{Ga}) of 7 nm/min and an N-flux (Φ_N) of 5 nm/min. The coalescence of the GaN islands formed in the early stage of the growth can clearly be seen. Figure 1b shows the top view image of self-assembled GaN NRs grown at 800°C with Φ_{Ga} = 4.3 nm/min and Φ_N = 14 nm/min. It can be seen that only partial coalescence of some NRs takes place. Figure 1c shows the cross-section SEM image of sample in Figure 1b where NRs have an average height of 1.2 μm and are

40 nm in diameter. Low temperature PL spectra of GaN samples shown in Figure 2 reveal strong differences depending on crystal quality. For NRs the spectrum is dominated by the D^0X line at 3.47 eV and the TES (Two-Electron Satellite) at around 3.45 eV. The line width of the D^0X is around 4-5 meV indicating a high "optical quality" of the NRs. In contrast PL from GaN layer shows a clear peak at 3.45 eV with a line width of 44 meV, revealing both the presence of tensile strain and a much lower material quality.

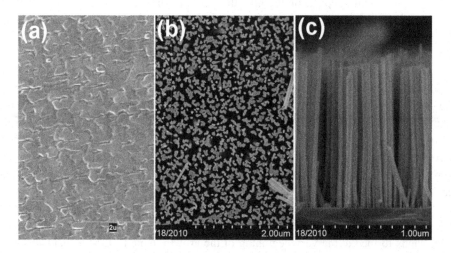

Fig. 1. SEM images of a) a GaN layer (top view), b) GaN NRs (top view), and GaN NRs (cross-section).

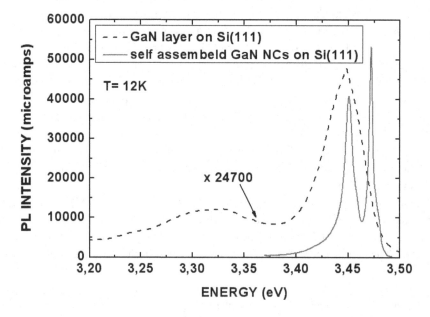

Fig. 2. PL spectra of a GaN layer (dashed line) and GaN NRs (solid line).

3.2. *Growth of self-assembled InGaN/GaN nanorods on Si(111) substrates*

A successful growth of InGaN/GaN nanocolumnar structures needs a previous understanding of the mechanisms that control the incorporation of In,[14] that mainly depends on: (i) In and Ga fluxes, (ii) active nitrogen amount, and (iii) growth temperature. Points (i) and (ii) reflect the fact that the Ga affinity to bind a nitrogen atom forming GaN is higher than that of In to form InN. Thus, a high In/Ga ratio as well as a high amount of active nitrogen would favor the incorporation of In. Figure 3 shows a clear PL peak shift from 2.56 eV to 2.44 eV when increasing the active nitrogen amount while keeping all other growth parameters fixed.

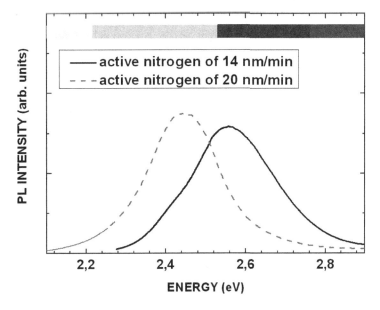

Fig. 3. PL spectra of InGaN NRs grown with different active nitrogen amounts.

Among the three variables mentioned before, the substrate temperature is by far the most important to successfully grow InGaN NRs since it affects the Ga and In desorption and diffusion processes as well as the decomposition of GaN and InN. It is important to note that the desorption of In and the decomposition of InN are significantly higher than the corresponding to Ga and GaN at the typical temperatures to grow self-assembled GaN NRs.[25,33] Then the approach is to reduce the growth temperature for InGaN NRs to values compatible with the InN decomposition and In desorption rates. In order to study this effect of the growth temperature, a series of InGaN NCs were grown following a two-step method: (i) 1 hour growth of GaN NCs at 800°C, with Φ_{Ga} = 4.3 nm/min and Φ_N = 14 nm/min; and (ii) 1 hour growth of InGaN NCs on top of the GaN NCs, at different fixed temperatures (650°C, 675°C, 700°C, 725°C and 750°C), keeping the same previous Φ_{Ga}, Φ_N values and $\Phi_{In} = \Phi_{Ga}$. In order to estimate effects of the InGaN growth on the NR morphology a reference sample of GaN NRs was grown for two hours under the same conditions as the GaN part of the InGaN/GaN NRs.

Figure 4 shows the In incorporation in these InGaN/GaN NRs as a function of the temperature derived from PL measurements. It can clearly be seen that for lower growth temperatures the In incorporation is increased up to a point of saturation (around 675°C), i.e. all the In supplied is incorporated. The In content can be estimated from the InGaN PL peak energy position according to Wu et al.,[34] being ~15% at 725°C; ~22% at 700°C and 33% at 675°C and 650°C.

As pointed out at the beginning of this section, diffusion and desorption are important variables which are strongly affected by the temperature. As a consequence significant changes in the NRs morphology should be expected as a function of the growth temperature. Indeed, two distinct effects are observed when lowering the growth temperature: (i) a change of growth rate, and (ii) an increase in the NR diameter.

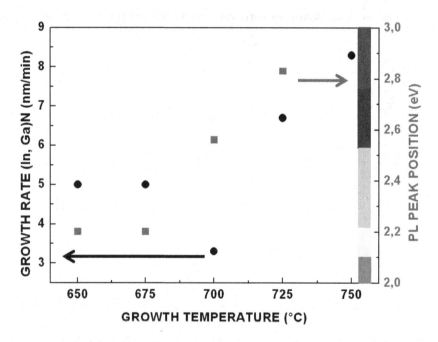

Fig. 4. Low temperature PL peak position and FWHM of InGaN/GaN NRs grown at different temperatures (red squares). Black circles show the growth rate dependence of the InGaN section of NRs on temperature. Reprinted with permission from S. Albert, A. Bengoechea-Encabo, P. Lefebvre, M. A. Sanchez-Garcia, E. Calleja, U. Jahn, A. Trampert, Emission control of InGaN nanocolumns grown by molecular-beam epitaxy on Si(111) substrates, *Appl. Phys. Lett.* **99**, 131108 (2011).

When neglecting the nucleation time, the growth rate of the GaN NRs reference sample can be estimated from Figure 5a to be around 10 nm/min. As shown in Figure 5b, for the InGaN NRs grown at 700°C a decrease in total NR length (1.0 versus 1.2 µm) and an increase in diameter (100 nm versus 50 nm) as compared to the reference sample can

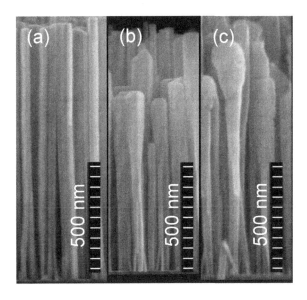

Fig. 5. SEM pictures of: (a) GaN NCs reference sample (growth at 800°C); (b) and (c) are selected InGaN/GaN NCs samples of series A with InGaN grown at 700°C and 650°C. Reprinted with permission from S. Albert, A. Bengoechea-Encabo, P. Lefebvre, M. A. Sanchez-Garcia, E. Calleja, U. Jahn, A. Trampert, Emission control of InGaN nanocolumns grown by molecular-beam epitaxy on Si(111) substrates, *Appl. Phys. Lett.* **99**, 131108 (2011).

be observed. From cross sectional SEM the growth rate of the InGaN part of the InGaN/GaN NRs can directly estimated for each growth temperature.

These values are plotted in Figure 4, indicating an initial decrease in the InGaN growth rate when lowering the growth temperature from 750 to 700°C, and a partial recovery for temperatures below 700°C. The total metal flux ($\Phi_{Ga} + \Phi_{In}$) being present during the growth of the InGaN NR part can be considered to be close to Φ_{Ga} in the temperature range between 750 and 700°C due to the very strong In desorption, as indicated by the PL peak position (Figure 4).[25] With that in mind the reduction of the Ga-adatoms diffusion length when lowering the sample temperature may account for the observed reduction in growth rate. Please note that it has been shown previously that the growth rate of GaN NRs is limited by the Ga-flux and that the diffusion of Ga-ad-atoms up the NR sidewalls is the main Ga source for the self-assembled GaN NR growth.[24,30]

Based on these arguments the partial recovery of the growth rate shown in Figure 4 can be explained by an enhanced incorporation of In due to a lower In desorption and InN decomposition.

The increase in the diameter of the NRs is observed to take place below the nominal GaN/InGaN interface, with the NR gradually widening up to the top (Figure 4b). From cathodoluminescence date (not shown) it can be seen the emissions related to InGaN arise only from material above this interface pointing towards Ga ad-atoms as the responsible for the diameter increase below the nominal interface. This can be explained by a longer residence time of Ga ad-atoms on the NR sidewall increasing the probability for them to bind with nitrogen before desorbing or reaching the NR top. Above the nominal GaN/InGaN NR interface both In and Ga will contribute to the lateral growth. When decreasing the growth temperature further the diameter increase can lead to the coalescence of the NRs (Figure 4c).

As discussed before, self-assembled InGaN NRs show in general a polychromatic PL emission mainly due to fluctuations of In%. This fact, together with inhomogeneities on the current injection (electroluminescence), the appearance of black spots due to defects formation upon NRs merging, and the difficulty to process them in a planar fashion, prevents a commercial exploitation for LEDs, particularly for monochromatic ones.[15,16] However, self-assembled InGaN NRs are quite helpful to study different strategies to generate white light. Since the growth temperature is one of the main factors that control the In incorporation, a temperature gradient during the InGaN NR growth should lead to a broad emission (white).

Figure 6 shows PL emission from an InGaN/GaN NRs array (self-assembled) where the InGaN region was graded in composition by changing the temperature from 750°C to 650°C, leading to an approximated change in average In% from 0 to 33% (Figure 4). PL emission at low temperature is indeed broad (white) and rather intense. However, the PL intensity quenches significantly at room temperature, being this effect much stronger on the "high energy" side of the spectrum.

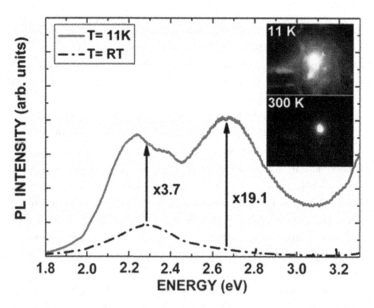

Fig. 6. Room- and low temperature PL of InGaN/GaN NRs grown with a temperature gradient. Reprinted with permission from S. Albert, A. Bengoechea-Encabo, P. Lefebvre, M. A. Sanchez-Garcia, E. Calleja, U. Jahn, A. Trampert, Emission control of InGaN nanocolumns grown by molecular-beam epitaxy on Si(111) substrates, *Appl. Phys. Lett.* **99**, 131108 (2011).

This result may be explained by the diffusion of carriers from regions of lower in content (lower localization energy) towards regions of higher In content [17].

3.3. *Growth of InGaN single quantum disks embedded into self-assembled GaN NRs*

Other approach to increase the efficiency of InGaN/GaN self-assembled nanocolumnar heterostructures for light emission is the use of InGaN QDiscs embedded into a GaN NR, leading to an increase in oscillator strength due to a higher confinement. The InGaN QDisc chosen for the confinement study was grown under the same conditions, in terms of absolute III/V ration and In/Ga ratio, as the InGaN NRs discussed in the previous section. Having in mind that the InGaN NRs already showed a complete incorporation of supplied In when grown at 675°C, and assuming a similar behavior when growing QDiscs, these last were grown at 625°C while all other conditions were kept constant a rough estimation of the In% in the QDisc, without considering possible strain related effects, can be derived from PL data, being around 33%, as derived from Figure 4.[34]

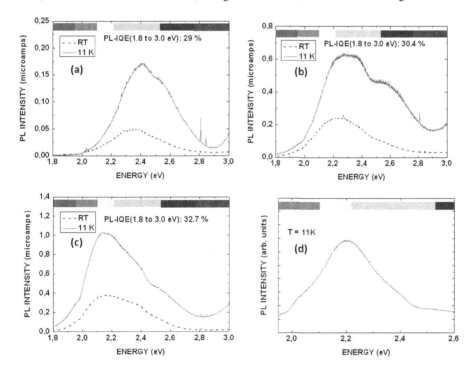

Fig. 7. Room- low temperature PL spectra of single InGaN quantum disk with a nominal thickness of a) 4 nm, b) 8 nm, and c) 16 nm in comparison to d) low temperature PL spectra of bulk InGaN/GaN NR with InGaN grown at 675°C.

A series of 3 samples with nominal QDisc thicknesses of 4 nm, 8 nm, and 16 nm was characterized by PL. Figure 6 shows the PL spectra of these samples at RT and 11 K in comparison to a bulk InGaN/GaN NR sample with the InGaN part grown at 675°C. For all three QDisk samples an InGaN related PL emission consisting of several peaks can be observed. As expected in terms of quantum confinement a blue shift of the overall InGaN related PL emission compared to the InGaN/GaN NR sample (Figure 6d) can be observed for a nominal QDisk thickness of 4 nm and 8 nm. For the nominally 16 nm thick QDisk no confinement is visible. For all three samples PL internal quantum efficiencies (PL-IQE) of around 30% are reached.

3.4. *Use of colloidal lithography to generate nanohole masks*

In order to achieve SAG and fabricate ordered arrays of NRs, a nanohole mask must be fabricated on the GaN/sapphire templates, generally obtained by e-beam lithography (EBL). Another technique that provides cheap, fast and almost periodical geometries is the colloidal lithography. In this section a discussion of both masking approaches is presented with special focus on the colloidal lithography. The mask material can be either metal (Ti, Mo) or dielectric (SiO_2, Si_3N_4), but this study will focus on Ti masks

In EBL the mask manufacture consists of a thin Ti film deposition, e-beam patterning of ordered arrays of nanoholes, and transfer of the pattern to the mask underneath by dry etching. To obtain optimal results, additional cleaning steps have to be performed after each of these steps. The most critical aspects that affect the mask surface are roughness, cleanliness, and adhesion of the Ti film to the GaN substrate. When all these parameters are optimized, high quality nanomasks are obtained yielding to a successful and reproducible ordered nanocolumnar growth. These aspects have been thoroughly discussed elsewhere.[35]

Besides EBL, colloidal lithography was used to fabricate Ti masks. As already mentioned, colloidal lithography is an easy, fast and cheap technique compared with others.[36,37] Though not having a perfect periodicity, these masks are easily fabricated over wide areas (2 inch or more) in a short period of time, as compared to EBL, and they allow the study of the SAG process and optimization that will be then transferred to a more standard patterning technique like Nanoimprint Lithography (NIL).

Colloidal lithography is based on the deposition of nanobeads on the GaN surface to be masked by Ti. First the GaN surface is pre treated by a sequential immersion in solutions of poly-electrolytes negative (PSS) and positively-charged (PDDA). The result is a trilayer PSS/PDDA/PSS, a few nanometers thick, covering the GaN surface. The negatively charged colloidal nanobeads used are made of polystyrene (PS) with diameters around 270 nm.

The solution (4% in weight) is deposited on the pre-treated GaN surface by spinning. As a result, areas with one monolayer of hexagonal close-packed nanobeads on the surface (scheme in Figure 7a, and example by SEM picture in Figure 7e) are achieved. A following oxygen plasma treatment reduces the diameter of the nanobeads, while keeping the distance between them (pitch) constant (Figure 7b). Then, 7 nm of Ti are thermally evaporated on the decorated surface (Figure 7c). Finally, nanobeads are removed from the surface, by tape stripping and organics based cleaning, leading to a GaN surface masked by a 7 nm Ti layer with nanoholes (scheme in Figure 7d and SEM picture of a real mask on GaN in Figure 7f).

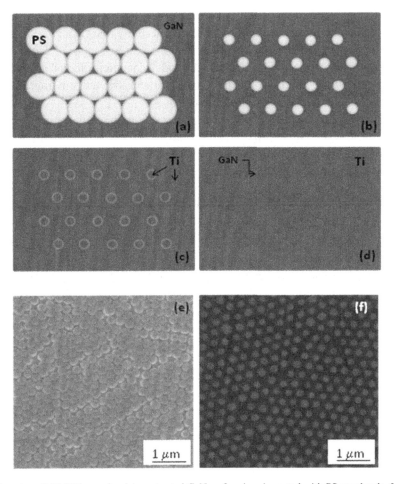

Fig. 8. Steps in colloidal lithography. (a) pre-treated GaN surface is spin coated with PS nanobeads, forming a monolayer with hexagonal symmetry. (b) PS/GaN is treated by oxygen plasma in order to reduce the diameter of nanobeads, and separate them from each other. (c) Ti metallization and (d) removal of the nanobeads (lift-off), leading to the Ti mask on GaN. Figures (e) and (f) show SEM pictures of a hexagonal close-packed PS nanobeads monolayer on GaN, and the resulting Ti mask on GaN, respectively.

3.5. *Growth of ordered NRs on c-plane and a-plane GaN/sapphire substrates*

SAG of GaN NRs needs to achieve simultaneously: (i) selectivity, and (ii) onset of NRs on each nanohole. Since both steps are interconnected and dependent on the selected growth conditions (temperature, III/V) as well as on the mask geometry and material used, it seems plausible that the optimal growth conditions for SAG may differ considerably from those applicable to self-assembled growth. Indeed, these differences are mainly a much higher growth temperature and a very high Ga flux to compensate a huge desorption from the substrate surface. This can be understood based on the requirements to fulfill selectivity.

Selectivity means that no growth must take place on the mask, but only within the nanoholes defined by the mask. In order to reach selectivity, the Ga atoms impinging on the mask must either desorb from it, or diffuse to the nearest nanohole, that is, avoiding nucleating GaN on the mask. This explains the "odd" conditions needed as compared to the self assembled growth. Both a high temperature and low active nitrogen act in the same way when thinking in terms of selectivity.

A high temperature is needed in order to increase both the diffusion length of Ga on the mask and also the Ga desorption rate from it to reduce the probability to nucleate GaN on it. This means a delicate balance because a too high temperature will certainly avoid nucleation on the mask by the strong Ga desorption, but it may also hinder any growth within the nanoholes. On the other hand, for a given temperature, selectivity will be enhanced for a lower amount of active nitrogen because the probability for Ga atoms to bind a N atom while on the mask are reduced. Again this means a balance because a low nitrogen amount may considerably reduce the NRs growth rate within the nanoholes.

Finally, the mask geometry is also a factor that must be considered for optimal SAG. The reason being that for a given temperature and nitrogen amount the distance between nanoholes (pitch), if too long, may favor GaN nucleation on the mask. This will happen if the pitch is too long in respect to the Ga atoms diffusion length (that depends on temperature). In addition, the size of nanoholes (diameter) for a given pitch will change the total density of Ga atoms within them, because of the Ga arriving by diffusion along the mask.

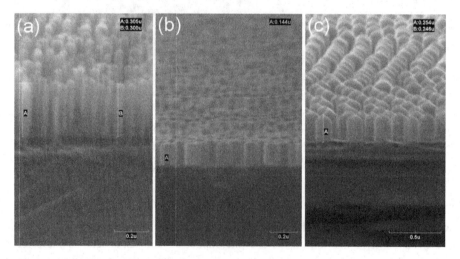

Fig. 9. SEM images of orderd GaN NRs with different diameters: a) 50 nm, b) 100 nm, and c) 150 nm.

Having all these factors in mind, ordered NRs of diameters from few tens of nanometers to almost one micron can be grown selectively.[19,20,21,22] In this work, ordered arrays of GaN NRs with diameters ranging from 50nm to 150nm were successfully grown, as shown in Figure 9, using colloidal lithography. Note that different morphologies of the NR top surface can be distinguished. Namely, pyramidal "pencil like" (Figure 9c) or flat (Figure 9b) can be achieved depending on the growth conditions. A detailed discussion of this aspect is going to follow in an upcoming work, though it can be said that the main factor that determines the NR top profile is the density of Ga atoms within the nanohole at the early nucleation stage.

Figure 10 shows SEM images (side and top) of an array of GaN NRs grown, in this case, on a EBL Ti mask. The dispersions in height and diameter with relative standard deviation of 1.6% and 2.9% respectively are quite small compared to those observed in self-assembled GaN NRs with standard deviations of 25%-60% in height and 10%-45% in diameter.[24,38,39]

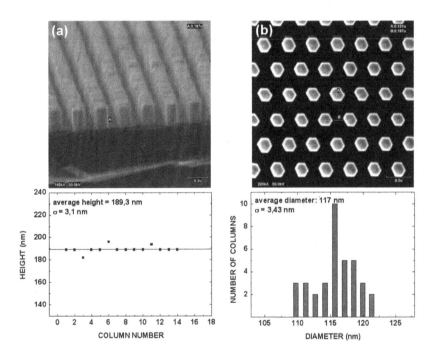

Fig. 10. SEM images of an array of ordered GaN NRs and the corresponding height and diameter distributions: a) side view and height distribution, b) top view and diameter distribution. This array was grown on a EBL Ti mask.

The cross-sectional scanning TEM bright-field image in Figure 10a shows that the diameter of the NR agrees well with the nanohole size defined on the Ti mask. The dark contrast lines at the interface between NR and GaN template indicate structural imperfections due to basal plane stacking faults (SF). The presence of stacking faults in the NRs is confirmed by the diffraction contrast image shown in Figure 10b. Besides these SFs no other extended defects, like threading dislocations, are observed. This is proven by the lattice images in Figures 10c and 10d showing that most of the stacking faults cross the whole NR and therefore do not lead to the formation of partial dislocations. Since in SAG of GaN NRs no epitaxial strain is involved (homoepitaxy on GaN templates) these stacking faults could be attributed to impurities such as titanium.

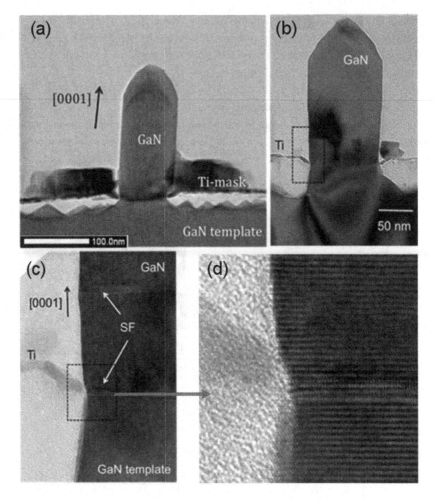

Fig. 11. a) cross-sectional scanning TEM bright-field image, b) cross-sectional scanning TEM diffraction contrast image, and c, d) cross-sectional scanning TEM lattice images of an ordered GaN NR.

In order to evaluate the optical quality of the ordered NRs, PL spectra have been taken at low temperature and compared to those from self assembled NRs, as shown in Figure 11.

Both the self assembled and the ordered NRs show a peak at around 3.47 eV which can be attributed to the D^0X indicating that in both cases the NRs, as expected, are virtually strain free.[4,32] The line widths of the D^0X line for both ordered and self assembled NRs are around 5 meV.

Another advantage of SAG to be exploited for broad light emission relies on the control to terminate GaN NRs, either with flat tops, or pyramidal ones. The growth of InGaN QDiscs in such NRs will follow the overall growth front geometry on the GaN NCs, which will lead to QDisc shapes varying from cylindrical shapes to more

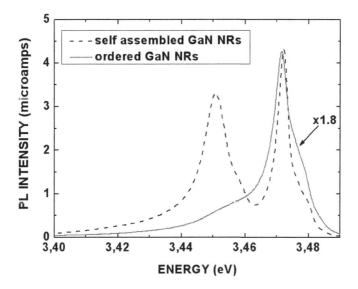

Fig. 12. Normalized PL spectra of ordered NRs (solid line) and self assembled NCs (dashed line) at 11K.

complicated ones involving different planes (polar, non-polar and semi-polar). Since the growth on non-polar or semi-polar planes may reduce or suppress the internal electric field, the problems caused by quantum confined Stark effect may also be reduced or entirely suppressed.[40]

In the case of SAG of GaN NRs on c-plane GaN templates, the NR growth front, though controllable as mentioned before, is in most cases formed by semi-polar planes (r-planes) and c-plane on top that yields a "pencil-like" profile.[19,20,21,22] When growing an InGaN QDisc on them, a similar or even more complicated profile is generated for the QDisc.[21]

As mentioned before, this kind of heterostructure may improve the light emission broadening (white) as well as its intensity due to a stronger electron-hole wave-functions overlapping. However, another, possibly better, approach is to grow the InGaN QDiscs and GaN NRs directly on non-polar GaN [10-10] (*m*-planes) or [11-20] (*a*-planes) planes.

The growth of ordered GaN NRs on *a*-plane GaN/sapphire templates by PAMBE has recently been demonstrated.[37] The same conditions and critical steps are applicable for the SAG of GaN NRs either on *c*-planes or *a*-planes. Perfect selectivity on colloidal lithography masks is achieved on *a*-plane GaN/sapphire templates, as shown in Figure 12. However, in this case, significant differences concerning the lateral and vertical growth rates can be seen.

For SAG of GaN NRs on a *c*-plane GaN template, hexagonal prisms ending with a pyramidal or flat top are obtained. In the case of SAG on *a*-plane GaN templates, elongated (lateral) nanostructures with different facets are obtained, being the top most one an *m*-plane, as shown in Figure 13. The direction along which these nanostructures show a pyramidal lateral termination (Figure 12a) corresponds to the *c*-plane. The cross section on Figure 13 was cut along this direction that shows nanostructure elongation.

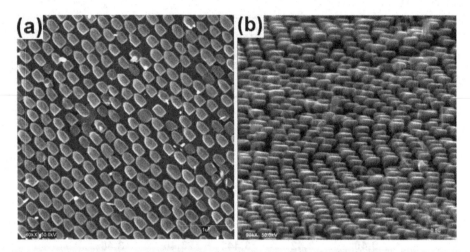

Fig. 13. SEM image of ordered GaN NRs on *a*-plane GaN template: a) top view. b) bird's eye view. a) Reprinted with permission from A. Bengoechea-Encabo, S. Albert, M.A. Sanchez-Garcia, L.L. López, S. Estradé, J.M. Rebled, F. Peiró, G. Nataf, P. de Mierry, J. Zuniga-Perez, E. Calleja, Selective area growth of *a*- and *c*-plane GaN nanocolumns by molecular beam epitaxy using colloidal nanolithography, *J. Crystal Growth* **353**, 1, 1-4 (2012).

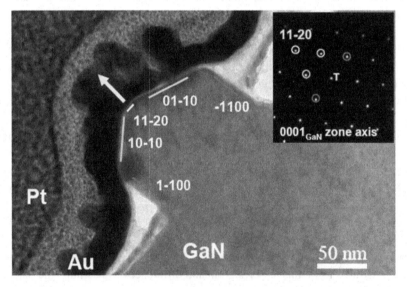

Fig. 14. View of the NR exhibiting facets along planes perpendicular to corresponding g vectors for most of the spots in 0001 axis diffraction pattern (inset). Reprinted with permission from A. Bengoechea-Encabo, S. Albert, M.A. Sanchez-Garcia, L.L. López, S. Estradé, J.M. Rebled, F. Peiró, G. Nataf, P. de Mierry, J. Zuniga-Perez, E. Calleja, Selective area growth of *a*- and *c*-plane GaN nanocolumns by molecular beam epitaxy using colloidal nanolithography, *J. Crystal Growth* **353**, 1, 1-4 (2012).

The vertical growth direction is confirmed to be [11-20], perpendicular to the *a*-plane. The nanostructures exhibits lateral faceting perpendicular to [1-100] and top facets perpendicular to [10-10] and [01-10], i.e. m-planes. The corresponding selected area diffraction pattern along the [0001] axis is shown in the inset of Figure 13, indicating that the nanostructure elongation occurs in that direction. Same results have been observed for similar structures grown by MOCVD.[41]

Figure 14 shows the measured vertical and lateral growth rates for GaN NRs grown on *c*-planes and *a*-planes. The growth on c-planes is quite anisotropic, that is, the axial growth rate dominates by far, what leads to well developed NRs with a high aspect ratio. On the other hand, while the NRs lateral growth on *c*-plane is almost zero, it reaches a considerable value when growing on *a*-planes (ratio of 0.4). This ratio is kept almost constant when the active nitrogen is increased.

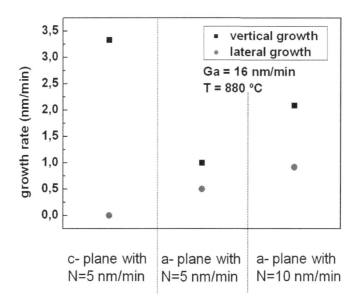

Fig. 15. Vertical and lateral growth rates for GaN NRs SAG on *c*-plane (growth temperature 880°C, $\Phi_{Ga} = 16$ nm/min, $\Phi_{N} = 5$ nm/min), *a*-plane under the same growth conditions, and *a*-plane under twice as much active nitrogen amount (10 nm/min). Reprinted with permission from A. Bengoechea-Encabo, S. Albert, M.A. Sanchez-Garcia, L.L. López, S. Estradé, J.M. Rebled, F. Peiró, G. Nataf, P. de Mierry, J. Zuniga-Perez, E. Calleja, Selective area growth of *a*- and *c*-plane GaN nanocolumns by molecular beam epitaxy using colloidal nanolithography, *J. Crystal Growth* **353**, 1, 1-4 (2012).

In order to evaluate the quality of non-polar SAG GaN NRs, PL spectra have been taken on them, as well as on the *a*-plane GaN template (prior to growth). Figure 15a clearly shows a strong yellow band and a rather weak emission at around 3.4 eV for the GaN template, indicating a rather low quality. On the other hand, the GaN NRs show no yellow band and a strong band edge emission with two peaks at around 3.36 and 3.42 eV. In Figure 15b PL spectra of SAG NRs on *a*-plane and *c*-plane are compared. On *c*-plane, GaN NRs are fully relaxed, and the dominant excitonic emission corresponds to the $D^{0}X$

line (Figure 11). However, on *a*-plane the dominant line is around 3.42 eV, that is red-shifted 50 meV in respect to the relaxed D^{0}X line. This may indicate a strong biaxial tensile strain for GaN NRs grown on *a*-plane. Actually, the band edge PL line from the GaN template (3.43 eV in Figure 15a) also corresponds to a red-shifted emission most likely due to biaxial tensile strain. It seems that there is a correlation between the two strain states in the template and the grown NRs, something that has never been observed when GaN NRs are grown on *c*-planes. The fact that the GaN NRs grown on *a*-plane are elongated and have a higher footprint as compared to their counterparts grown on *c*-planes, as well as a much smaller aspect ratio, maybe determinant to keep them under the biaxial tensile strain "inherited" from the template. In other words, a much higher nanostructure may eventually relax. A careful look at the NRs interface with the template by High Resolution TEM, checking for dislocation networks, may shine some light to better interpret these results.

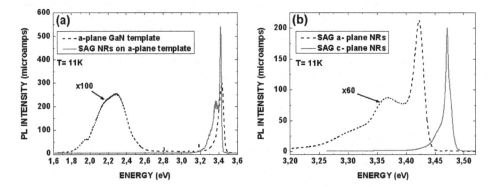

Fig. 16. PL spectra of a) *a*-plane GaN template, b) SAG GaN NRs on the *a*-plane GaN template.

3.6. *Growth of InGaN single quantum disks embedded into ordered GaN NRs*

As in the case of self-assembled NRs, embedded InGaN QDiscs can be grown embedded into ordered NRs, either on *c*- plane, or *a*-plane. Figure 16 shows the morphology and optical properties of these structures.

The PL spectrum of the InGaN QDiscs grown on *a*-plane SAG NRs shows a dominant emission line at 2.34 eV, attributed to the QDisc. The broad shoulder in the range of 2.6 eV to 3.3 eV is most probably related to the growth of InGaN on the TiN mask, as can be seen in Figure 16a (grey contrast between the NRs). The SEM image in Figure 16b reveal that the structures grown on *c*-plane GaN/sapphire templateshave a pyramidal top. From the PL spectrume two InGaN related peaks at around 2.45 eV and 2.95 eV are observed. The existence of two emission lines in the PL spectrum of InGaN QDiscs embedded on NRs with pyramidal top has already been reported elsewhere.[21] The low energy peak is thought to originate from the semi-polar side facet of the pyramidal top, while the high energy peak relates most likely to a polar plane developing at the apex formed by the convergence of six non-polar facets.

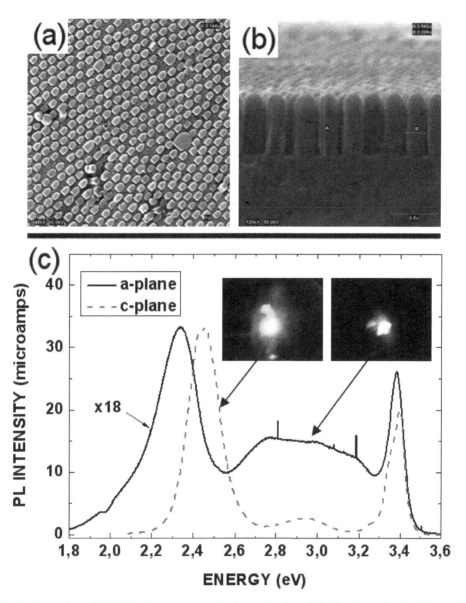

Fig. 17. Comparison of InGaN QDiscs grown on ordered *c*- and *a*-plane GaN NRs: a) top view SEM image of InGaN/GaN NRs grown on *a*-plane. b) side view SEM image of InGaN/GaN NRs grown on *c*-plane. c) Room temperature PL spectra of *a*- and *c*-plane InGaN/GaN NRs.

Very high IQE was observed in the case of PL emission from the InGaN nano-disk embedded on a-plane GaN nanostructures. Figure 18 shows the LT (12 K) and RT PL spectra comparison. The IQE of the overall spectra is 7%, but as it was explained before, the emission has different InGaN contributions. Although the emission coming from the InGaN grown on the Ti mask (range between 2.3 eV to 3.3 eV) has a relatively low IQE

(4.7%), as expected, the one coming from the embedded InGaN on the a-plane nanostructures, corresponding approximately to the range from 1.6 eV to 2.3 eV (red to green), has a IQE as high as 58%. Further improvement in the selective growth on a-plane templates promises an extraordinary effect on the performance of future devices.

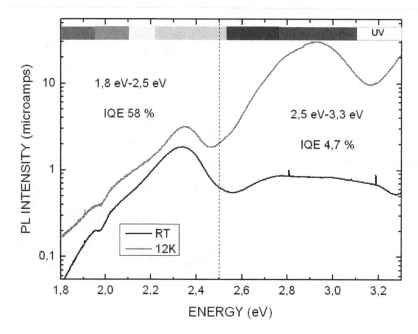

Fig. 18. LT and RT PL spectra of the InGaN nanodisk embedded on a-plane GaN nanostructures. Two different regions are distinguished. The emission at lower energy corresponds to the embedded InGaN nanodisk, with a very high IQE, up to 58%. The emission at higher energies, with lower IQE around 4.7%, corresponds to the InGaN non-selectively grown on the mask.

4. Summary

The first part of this work described basic principles involved in the growth of GaN self-assembled NRs grown on Si(111) substrates by PAMBE. Depending on the III/V ratio and sample temperature either GaN layers or GaN self-assembled GaN nanocolumns grow directly on Si(111). A significant improvement in optical quality of the self-assembled GaN NRs compared to GaN layers was observed.

Following that, the effect of the growth temperature on the morphology and emission characteristics of self-assembled InGaN NRs was studied. Within the growth temperature range of 650°C to 750°C, an increase in the In incorporation for decreasing temperature was observed. This effect allowed tailoring the InGaN NRs emission line shape by using temperature gradients during growth yielding white emission at room temperature and low temperature. Besides that the implementation of single InGaN QDisk structures into self-assembled GaN NRs was shown, revealing a clear blue-shift with decreasing nominal thickness of the quantum disk.

The second part of this work described the influence of the substrate temperature, and III/V flux ratio on the SAG of GaN NRs on both c-plane sapphire/GaN- and a-plane sapphire/GaN substrates. Two ways of lithography were presented putting special emphasis on the colloidal lithography, as a method which is able to handle big areas in a short time. With optimized conditions, ordered GaN NRs were grown with a wide variety of diameters. SAG of a-plane GaN NRs was compared, in terms of lateral, vertical growth rates, and optical properties with the SAG of c-plane GaN NRs. The vertical growth rate of ordered a-plane and c-plane nanocolumns was found to differ by a factor of 3.5, with the same conditions being used. Besides that it was shown that ordered a-plane NRs show a significant lateral growth rate, which is in contrast to the ordered c-plane NRs. For the optical properties it has been found that ordered c-plane NRs are fully relaxed and show a full width half maximum comparable to self-assembled GaN NRs. In the case of ordered a-plane NRs only partial relaxation with respect to the strained a-plane substrate and a higher full width half maximum compared to ordered c-plane GaN nanocolumns was observed. In addition to that the implementation of single InGaN QDisk structures into both ordered a-plane GaN and c-plane GaN NRs was shown revealing green emission for the single quantum disk implemented on ordered c-plane GaN NRs and white emission for the single QDisk implemented in ordered a-plane NRs at room temperature.

Based on the presented results, two lines are being under development in our group. First, following the RGB approach initiated with self-assembled NRs, the ordered growth of c-oriented InGaN NRs achieving red, green and blue emission, and their combination for white light emission. And second, the improvement of selectivity during the InGaN growth on a-plane nanostructures, since first results have shown a very high IQE (58%) for the InGaN embedded in such nanostructures.

Acknowledgments

We acknowledge partial financial support by the EU FP7 Contract SMASH 228999-2; the Initial training network RAINBOW project PITN-GA-2008-213238 and by Spanish projects CAM/P2009/ESP-1503 and MICINN MAT2008-04815.

References

1. S. Nakamura, G. Fasol, S.J. Pearton, *The Blue Laser Diode: The Complete Story* (Springer, Berlin, 2000).
2. F.K. Yam, Z. Hassan, InGaN: An overview of the growth kinetics, physical properties and emission mechanisms, *Superlattices & Microstructures* **43**, 1 (2008).
3. T. Kuykendall, P. Ulrich, S. Aloni, P. Yang, Complete composition tunability of InGaN nanowires using a combinatorial approach, *Nat. Mater.* **6**, 952 (2007).
4. E. Calleja, M.A. Sachnchez-Garcia, F.J. Sanchez, F. Calle, F.B. Naranjo, E. Munoz, U. Jahn, K. Ploog, Luminescence properties and defects in GaN nanocolumns grown by molecular beam epitaxy, *Phys. Rev. B* **62**, 16826 (2000).

5. R. Meijers, T. Richter, R. Calarco, T. Stoica, H.P. Bochem, M. Marso, H. Lüth, GaN-nanowhiskers: MBE-growth conditions and optical properties, *J. Crystal Growth* **289**, 381 (2006).

6. N. Thillosen, K. Sebald, H. Hardtdegen, R. Meijers, R. Calarco, S. Montanari, N. Kaluza, J. Gutowski, H. Luth, The state of strain in single GaN nanocolumns as derived from micro-photoluminescence measurements, *Nano Letters* **6**, 704 (2006).

7. M. Yoshizawa, A. Kikuchi, M. Mori, N. Fujita, and K. Kishino, Growth of Self-Organized GaN Nanostructures on Al_2O_3 (0001) by RF-Radical Source Molecular Beam Epitaxy, *Jpn. J. Appl. Phys.* **36**, L459 (1997).

8. Y.S. Park, C.M. Park, D.J. Fu, T.W. Kang, J.E. Oh, Photoluminescence studies of GaN nanorods on Si (111) substrates grown by molecular-beam epitaxy, *Appl. Phys. Lett.* **85**, 5718 (2004).

9. H.-M. Kim, Y.-H. Cho, H. Lee, S.I. Kim, S.R. Ryu, D.Y. Kim, T.W. Kang, K.S. Chung, High Brightness light emitting diodes using dislocation-free indium gallium nitride/gallium nitride multiquantum-well nanorod arrays, *Nano Letters* **4**, 1059 (2004).

10. C.H. Chiu, T.C. Lu, H.W. Huang, C.F. Lai, C.C. Kao, J.T. Chu, C.C. Yu, H.C. Kuo, S.C. Wang, C.F. Lin, T.H. Hsieh, Fabrication of InGaN/GaN nanorod light-emitting diodes with self-assembled Ni metal islands, *Nanotechnology* **18**, 445201 (2007).

11. Y. Kawakami, S. Suzuki, A. Kaneta, M. Funato, A. Kikuchi, K. Kishino, Origin of high oscillator strength in green-emitting InGaN/GaN nanocolumns, *Appl. Phys. Lett.* **89**, 163124 (2006).

12. H.-Y. Chen, H.-W. Lin, C.-H. Shen, S. Gwo, Structure and photoluminescence properties of epitaxially oriented GaN nanorods grown on Si(111) by plasma-assisted molecular-beam epitaxy, *Appl. Phys. Lett.* **89**, 243105 (2006).

13. H.-S. Chen, D.-M. Yeh, Y.-C. Lu, C.-Y. Chen, C.-F. Huang, T.-Y. Tang, C.C. Yang, C.-S. Wu, C.-D. Chen, Strain relaxation and quantum confinement in InGaN/GaN nanoposts, *Nanotechnology* **17**, 1454 (2006).

14. S. Albert, A. Bengoechea-Encabo, P. Lefebvre, M. A. Sanchez-Garcia, E. Calleja, U. Jahn, A. Trampert, Emission control of InGaN nanocolumns grown by molecular-beam epitaxy on Si(111) substrates, *Appl. Phys. Lett.* **99**, 131108 (2011).

15. H. W. Lin, Y. J. Lu, H. Y. Chen, H. M. Lee, and S. Gwo, InGaN/GaN nanorod array white light-emitting diode, *Appl. Phys. Lett.* **97**, 073101, (2010).

16. A. L. Bavencove, G. Tourbot, J. Garcia, Y. Désières, P. Gilet, F. Levy, B. André, B. Gayral, B. Daudin and L. S. Dang, Submicrometre resolved optical characterization of green nanowire-based light emitting diodes, *Nanotechnology* **22**, 345705, (2011).

17. W. Guo, M. Zhang, A. Banerjee and P. Bhattacharya, Catalyst-Free InGaN/GaN Nanowire Light Emitting Diodes Grown on (001) Silicon by Molecular Beam Epitaxy, *Nano Letters* **10**, 9, 3355–3359 (2010).

18. K. Kishino, A. Kikuchi, H. Sekiguchi, and S. Ishizawa, InGaN/GaN nanocolumn LEDs emitting from blue to red, *Proc. SPIE* **6473**, 64730T (2007).

19. H. Sekiguchi, K. Kishino, A. Kikuchi, Ti-mask Selective-Area Growth of GaN by RF-Plasma-Assisted Molecular-Beam Epitaxy for Fabricating Regularly Arranged InGaN/GaN Nanocolumns, *Appl. Phys. Express* **1**, 124002 (2008).

20. K. Kishino, H. Sekiguchi, A. Kikuchi, Improved Ti-mask selective-area growth (SAG) by rf-plasma-assisted molecular beam epitaxy demonstrating extremely uniform GaN nanocolumn arrays, *J. Crystal Growth* **311**, 2063 (2009).

21. H. Sekiguchi, K. Kishino, A. Kikuchi, Emission color control from blue to red with nanocolumn diameter of InGaN/GaN nanocolumn arrays grown on same substrate, *Appl. Phys. Lett.* **96**, 231104 (2010).

22. A. Bengoechea-Encabo, F. Barbagini, S. Fernandez-Garrido, J. Grandal, J. Ristic, M.A. Sanchez-Garcia, E. Calleja, U. Jahn, E. Luna, A. Trampert, Understanding the selective area growth of GaN nanocolumns by MBE using Ti nanomasks, *J. Crystal Growth* **325**, 89 (2011).

23. B. Heying, R. Averbeck, L. F. Chen, E. Haus, H. Riechert, J.S. Speck, Control of GaN surface morphologies using plasma-assisted molecular beam epitaxy, *J. Appl. Phys.* **88**, 1855 (2000).

24. S. Fernández-Garrido, J. Grandal, E. Calleja, M.A. Sánchez-Garcia, and D. López-Romero, A growth diagram for plasma-assisted molecular beam epitaxy of GaN nanocolumns on Si(111), *J. Appl. Phys.* **106**, 126102 (2009).

25. C.S. Gallinat, G. Koblmüller, J.S. Brown, J.S. Speck, A growth diagram for plasma-assisted molecular beam epitaxy of In-face InN, *J. Appl. Phys.* **102**, 064907 (2007).

26. Y.S. Park, S.-H. Lee, J.-E. Oh, C.-M. Park, T.-W. Kang, Self-assembled GaN nano-rods grown directly on (1 1 1) Si substrates: Dependence on growth conditions, *J. Cryst. Growth* **282**, 313 (2005).

27. R. Calarco, R.J. Meijers, R.K. Debnath, T. Stoica, E. Sutter, H. Lüth, Nucleation and Growth of GaN Nanowires on Si(111) Performed by Molecular Beam Epitaxy, *Nano Letters* **7**, 2248 (2007).

28. R.K. Debnath, R. Meijers, T. Richter, T. Stoica, R. Calarco, H. Lüth, Mechanism of molecular beam epitaxy growth of GaN nanowires on Si(111), *Appl. Phys. Lett.* **90**, 123117 (2007).

29. O. Landré, R. Songmuang, J. Renard, E. Bellet-Amalric, H. Renevier, and B. Daudin, Plasma-assisted molecular beam epitaxy growth of GaN nanowires using indium-enhanced diffusion, *Appl. Phys. Lett.* **93**, 183109 (2008).

30. J. Ristic, E. Calleja, S. Fernández-Garrido, L. Cerutti, A. Trampert, U. Jahn, K.H. Ploog, On the mechanisms of spontaneous growth of III-nitride nanocolumns by plasma-assisted molecular beam epitaxy, *J. Crystal Growth* **310**, 4035 (2008).

31. C.T. Foxon, S.V. Novikov, J.L. Hall, R.P. Campion, D. Cherns, I. Griffiths, S. Khongphetsak, A complementary geometric model for the growth of GaN nanocolumns prepared by plasma-assisted molecular beam epitaxy, *J. Crystal Growth* **311**, 3423 (2009).

32. A. Trampert, J. Ristic, U. Jahn, E. Calleja, K.H. Ploog, TEM study of (Ga,Al)N nanocolumns and embedded GaN nanodiscs, *Inst. Phys. Conf. Ser.* **180**, 167 (2003).

33. S. Fernández-Garrido, G. Koblmüller, E. Calleja, J.S. Speck, *In situ* GaN decomposition analysis by quadrupole mass spectrometry and reflection high-energy electron diffraction, *J. Appl. Phys.* **104**, 033541 (2008).

34. J. Wu and W. Walukiewicz, Band gaps of InN and group III nitride alloys, *Superlattices & Microstructures* **34**, 63 (2003).

35. F. Barbagini, A. Bengoechea-Encabo, S. Albert, J. Martinez, M.A. Sanchez García, A. Trampert, E. Calleja, Critical aspects of substrate nanopatterning for the ordered growth of GaN nanocolumns, *Nanoscale Research Letters* **6**, 632 (2011).

36. J.C. Hulteen and R.P. Van Duyne, Nanosphere Lithography: A materials general fabrication process for periodic particle array surfaces, *J. Vac. Sci. Technol. A* **13**(3), 1553 (1995).

37. A. Bengoechea-Encabo, S. Albert, M.A. Sanchez-Garcia, L.L. López, S. Estradé, J.M. Rebled, F. Peiró, G. Nataf, P. de Mierry, J. Zuniga-Perez, E. Calleja, Selective area growth of *a*- and *c*-plane GaN nanocolumns by molecular beam epitaxy using colloidal nanolithography, *J. Crystal Growth* **353**, 1, 1-4 (2012).

38. M. Yoshizawa, A. Kikuchi, N. Fujita, K. Kushi, H. Sasamoto, K. Kishino, Self-organization of GaN/Al$_{0.18}$Ga$_{0.82}$N multi-layer nano-columns on (0 0 0 1) Al$_2$O$_3$ by RF molecular beam epitaxy for fabricating GaN quantum disks, *J. Crystal Growth*, **189-190**, 138 (1998).

39. H.-Y. Chen, H.-W. Lin, C.-H. Shen, S. Gwo, Structure and photoluminescence properties of epitaxially oriented GaN nanorods grown on Si(111) by plasma-assisted molecular-beam epitaxy, *Appl. Phys. Lett.* **89**, 243105 (2006).

40. P. Waltereit, O. Brandt, A. Trampert, H.T. Grahn, J. Menniger, M. Ramsteiner, M. Reiche, K.H. Ploog, Nitride semiconductors free of electrostatic fields for efficient white light-emitting diodes, *Nature* **406**, 865 (2000).
41. V. Jindal and F. Shahedipour-Sandvik, Theoretical prediction of GaN nanostructure equilibrium and nonequilibrium shapes, *J. Appl. Phys.* **106**, 083115 (2009).

CATALYST-FREE GaN NANOWIRES AS NANOSCALE
LIGHT EMITTERS

KRIS BERTNESS*, NORMAN SANFORD, JOHN SCHLAGER, ALEXANA ROSHKO,
TODD HARVEY, PAUL BLANCHARD, MATTHEW BRUBAKER,
ANDREW HERRERO and ARIC SANDERS

National Institute of Standards and Technology
Boulder, CO, USA
*bertness@boulder.nist.gov

Catalyst-free growth of GaN nanowires with molecular beam epitaxy produces material of exceptionally high quality with long minority-carrier lifetimes and low surface recombination velocity. The nanowires grow by thermodynamic driving forces that enhance the sticking coefficient of incoming reagents to the end facets of the nanowire while inhibiting growth on the *m*-plane sidewalls. Photoluminescence (PL) studies confirm that the material is essentially free of detrimental chemical impurities and crystalline defects. The nanowires are readily excited to lasing with modest optical pump power. Recent progress in methods for selective epitaxy has made it possible to control both the diameter and placement of the nanowires. Despite the high material quality, the energy-conversion efficiency of *single* nanowire LEDs remains low. The primary limitation appears to be optimizing the *p*-type doping with Mg, which is both a growth and a measurement problem.

Keywords: GaN nanowires; molecular beam epitaxy; Mg doping; optoelectronic devices.

1. Introduction

The breakthrough in GaN optoelectronics that occurred in the late 1980s is largely credited to the work of groups led by Amano [Amano et al., 1989] and Nakamura [Nakamura et al., 1992], who separately demonstrated active *p*-type doping through reduction of native defects and activation anneals that reverse hydrogen passivation of Mg dopant atoms. Despite the rapid growth of the GaN-based LED industry, the internal quantum efficiency of green LEDs remains below 25%. Recent improvements have focused heavily on semipolar and nonpolar growth schemes in order to mitigate the internal spontaneous and piezoelectric electric fields that form at heterojunctions. Such symmetry-dependent electrostatic artifacts can substantially reduce recombination efficiency in quantum wells. [Detchprohm et al., 2010] Additionally, control of *p*-type doping and defects is still under intense study. [Huang et al., 2011; Li et al., 2011; Zhang and Yao, 2011] Although the availability and quality of AlN and GaN substrates has improved significantly, they remain small (<75 mm diameter) and contain defect densities that are orders of magnitude higher than what is commonly achieved with the growth of Si, GaAs, and InP crystals. Most commercial GaN-based devices are grown on sapphire or SiC substrates, resulting in epitaxial GaN films with threading dislocation densities on the order of 10^6 to 10^{10} cm^{-2}.

Against this backdrop of successful industrial development of devices based on material with high defect densities, two groups [Yoshizawa et al., 1997; Sánchez-García et al., 1998] found that GaN spontaneously formed essentially defect-free nanowires when grown by molecular beam epitaxy (MBE). The MBE growth environment promotes high chemical purity, and the thermodynamic driving forces that cause the nanowires to form also encourage the movement of any dislocations formed during nanowire nucleation to the surface of the nanowire, where they terminate. As will be discussed further, catalyst-based growth of GaN nanowires does not necessarily lead to the same high-quality material, particularly when the growth environment and reactants contain contaminants. Although GaN nanowires should make excellent light emitting diodes (LEDs) and lasers, there remains a large gap between the promise of the materials properties and actual device performance. At the present time, we attribute this state to the absence of suitable surface passivation materials and to the need to reengineer doping and contacts for the nanowire growth process and morphology, particularly for *p*-type doping.

Because recent review articles [Calleja et al., 2007; Bertness et al., 2010] on MBE-grown GaN nanowires have been published, and because many of the major groups working in the field are represented in other papers in this proceedings, this paper will not describe the development of this specialty from a historical perspective nor will it present a comprehensive literature review. We will instead provide background material concerning what we presently understand about the growth of nanowires with MBE, and then analyze the challenges remaining for the production of efficient, single-nanowire LEDs. The requirements for efficient nanowire light emitters are (1) controlled morphology, (2) low defect density and long minority carrier lifetime, (3) *n*-type and *p*-type doping, (4) contact metallization for efficient carrier injection, and (5) control of surface depletion. As we will show, most of the remaining challenges lie in the *p*-type doping and contacting.

2. Growth Review

GaN nanowires form spontaneously during MBE growth under conditions of high N:Ga flux ratios and high substrate temperatures (800°C to 900°C). In MBE, Ga, In and Al are incorporated into the crystal from effusion cells containing liquid melts of the elements in high-purity form. The N source is typically molecular N_2 that has passed through a radio-frequency plasma zone, producing both atomic and excited molecular states that then have sufficient energy to react with the group III elements at the growing surface. The N species dominate the chamber pressure during growth, reaching values from ~10 mPa to ~ 50 mPa (8 x 10^{-6} Torr to 3 x 10^{-5} Torr). The most common *n*-type dopant is Si, also evaporated from effusion cells. The only successful *p*-type dopant for GaN is Mg, and this element poses particular difficulties for MBE because of its intermediate vapor pressure. The temperatures required to achieve the desired Mg flux are around 525 K to 625 K, over 500 K less than the operating temperatures of most other cells and the substrate itself. Our systems are now equipped with valved cracker sources for Mg, which helps produce a more stable beam. [Burnham et al., 2005]

Representative growth morphology is illustrated in Fig. 1(a). The nanowires will grow indefinitely at nearly uniform diameter provided that they do not coalesce. We typically run from one to five days to produce nanowires that are 10 to 20 μm long. The growth axis of the nanowires is along the [0001] direction, or *c*-axis, of the hexagonal (wurtzite) crystal structure that is the lowest-energy phase for GaN and related group III-nitride alloys. The sidewalls of the nanowires are one of the six equivalent <1 -1 0 0> planes or *m*-planes. Depending on the nucleation conditions and the growth habit, the sidewalls can be similar in width and remain so during growth to produce nanowires of nearly perfect hexagonal cross-section, or more irregular (but still hexagonal) forms may be seen. When grown on Si(111) substrates, the nanowires align with the underlying crystal structure of the Si and are thus aligned with each other. This registry allows x-ray diffraction analysis of nanowire ensembles, which has shown that the nanowires are strain-free. Although there is still a great deal to be learned about the nucleation process, the propagation of nanowire growth has been shown to be driven by differences in the sticking coefficient for incoming Ga atoms on the different crystal planes that form. [Bertness et al., 2007] The sticking coefficient is related to the surface energy and the availability of bonding sites, as well as surface diffusion coefficients, and naturally the surface temperatures must be high in order for any of the sticking coefficients to fall below unity and therefore become distinct among different surfaces. It has been shown definitively that the growth does not proceed via self-catalysis (i.e., Ga droplets are not present at the tips during growth). [Bertness et al., 2008; Ristić et al., 2008]

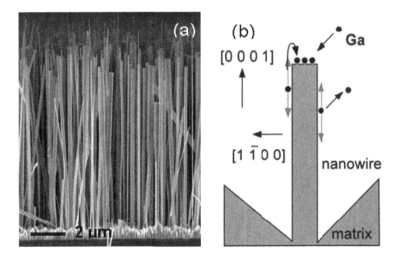

Fig. 1. (a) Typical morphology of GaN nanowires grown by MBE, and (b) schematic of growth mechanism.

3. Nucleation Processes

Nucleation is the least understood part of the growth process; however, both strain and morphology appear to affect nucleation. [Consonni et al., 2010] Our group has recently carried out experiments to examine the role of strain in nucleation with homoepitaxial growth, where strain was induced by scratching the GaN surface. The GaN template layer was marked in several places with scratches from a scribe, trenches made with different etching techniques, and scratches that were chemically etched after being scribed. Of these three markings, only the unaltered scratches are known to induce local strain from the nanoscale disruptions of the lattice. Etching processes tend to remove regions of high strain (around dislocations, for example) and thus can generate morphological changes without strain. As shown in Fig. 2, the strained scratches were the only areas where nanowires nucleated. Thus some degree of strain appears to be necessary under most circumstances to promote the formation of a nanowire seed crystal.

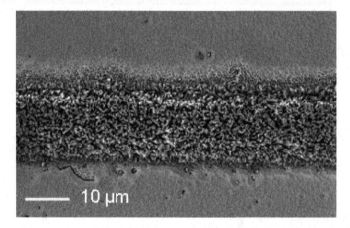

Fig. 2. Homoepitaxial growth of GaN nanowires on GaN, illustrating that strained regions produced by scoring the GaN surface allow nucleation, but undisturbed regions do not.

Morphology can also affect nucleation; in particular, nanowires tend to nucleate in small depressions in the surface. There are presently several methods of selective epitaxy for nucleating GaN nanowires in MBE. All make use of patterned mask layers, including TiN_x, [Kishino et al., 2009] SiO_x, [Calleja et al., 2007] and SiN_x. [Bertness et al., 2010] We have had the most success with the structure shown in Fig. 3(a). The mask is deposited on an MBE-grown AlN buffer layer similar to those used for random nucleation. The sticking coefficient on the SiN_x is less than that on the GaN, and over a narrow range in temperature (that also depends on Ga flux and N species), nearly perfect selectivity can be attained. Nanowires formed in these openings are very uniform in their diameter (variations $< \pm 3\%$), and those diameters are controlled by the diameter of the mask opening. [Bertness et al., 2010] The structures shown in Fig. 3(b), however, are slow-growing and sometimes composed of several parts (unlike those in the figure). By

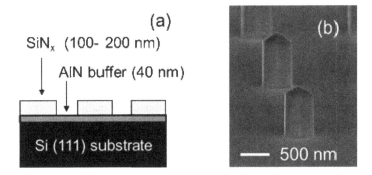

Fig. 3. (a) Selective epitaxy to control nucleation and location of nanowires. (b) GaN nanowires grown by use of the structure in (a).

Fig. 4. Morphology of GaN nanowires growing on AlN buffer layers with different distributions of Al-polar material. Al-polar domains are indicated by green overlay on the topographic image.

altering the V:III ratio during the nucleation process, longer, thinner nanowires form instead. The challenge remains to expand these nanowires to the larger diameters (>200 nm) needed for optical devices.

A third factor affecting nucleation is the polarity of the buffer layer crystals, [Brubaker et al., 2011] as shown in Fig. 4. For normal nanowire growth conditions, the polarity of the GaN growth follows that of the underlying AlN buffer layer (if present). Because polarity can affect the growth rate (more precisely, the decomposition rate under atomic N flux [VanMil et al., 2005]), it appears that growth of Ga-polar material occurs on and is defined by underlying Al-polar domains. Because AlN buffer layer polarity is a complex function of N:Al ratio, substrate temperature, and Al pre-layer types, the ability to measure this polarity with piezoresponse force microscopy has been an important growth characterization tool. [Brubaker et al., 2011]

4. Optical Properties for Optical Emitters

We frequently speak of GaN nanowires as being "defect-free." This statement is supported by structural studies such as transmission electron microscopy (TEM), but in fact the strongest evidence for the absence of defects comes from the optical characteristics of the nanowires. The low-temperature PL spectrum of MBE-grown GaN nanowires consists of free excitons, excitons bound to shallow donors, and phonon replicas of these peaks, as shown in Fig. 5. [Schlager et al., 2006] GaN nanowires grown with other methods frequently show additional peaks near the band edge or deeper into the visible (yellow and blue emission), indicating the presence of structural defects or chemical impurities. These extra PL peaks are sometimes seen in nanowire clusters that have coalesced during growth, just as similar clusters sometimes (though not always) contain chains of dislocations visible with TEM. Although narrow PL peaks are often equated with material purity, the low-temperature spectra of nanowires are frequently broadened by strain from bending during dispersal or strain induced by substrates with different thermal-expansion coefficients. Time-resolved PL is readily obtained for individual nanowires, with typical room-temperature lifetimes on the order of a few nanoseconds. The diameter dependence of the PL lifetime is linear if surface recombination dominates the nonradiative losses — as is the case at room temperature. The surface recombination velocity of our GaN nanowires at room temperature was thus determined to be 9×10^{-4} cm/s. [Schlager et al., 2008] This value is excellent for a bare semiconductor surface, but still requires improvement through some (yet undetermined) surface passivation method for optimum luminescence performance in optoelectronic devices. Our work has also shown that the PL lifetime has a complicated temperature dependence and intensity dependence. [Schlager et al., 2011] These effects invalidate the simple test that has become commonplace in the literature, which assumes that the luminescence at low temperature (about 10 K) necessarily results from unity internal quantum efficiency (IQE), and that the room-temperature IQE is equal to the PL intensity at room temperature divided by the PL intensity at low temperature. This ratio often gives an optimistically high number. Our more detailed study estimates that the IQE of a 460 nm diameter nanowire is about 30% at room temperature and with an excitation fluence of 190 μJ/cm^2.

MBE-grown nanowires lase readily under optical pumping, and this process reveals interesting facts about the nanowires as optical cavities, as shown in Fig. 6. The nanowires behave as optical waveguide lasers with extremely high numerical aperture. The light emitted from the end facets diverges so strongly that the ends of the nanowires act like point emitters with a resulting spatial interference pattern that can surround the wire. At the higher carrier densities needed for lasing, the losses that dominate are not the surface recombination paths but rather optical losses due to any taper in the nanowire diameter or irregular tip shapes. Figure 6(b) illustrates the spatial interference pattern produced by emission from the two end facets of a GaN nanowire laser. Similar images have been recorded for optically pumped ZnO nanowire lasers. [van Vugt et al., 2006]

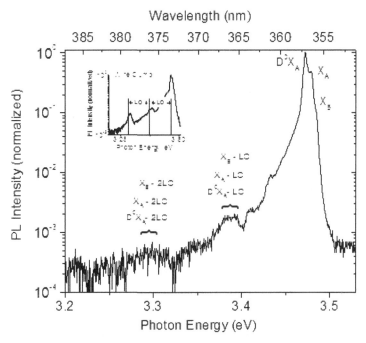

Fig. 5. Photoluminescence at 3.2 K for a single GaN nanowire and (inset) a small wire clump. Free excitons are noted with "X", donor-bound excitons as D^0X, and phonon replicas with the LO shift. The A and B subscripts refer to two different transitions in the GaN band structure.

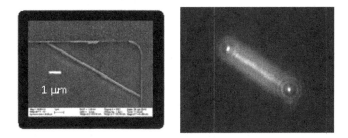

Fig. 6. SEM of dispersed GaN single nanowire and the same nanowire optically pumped to the lasing state. Typical thresholds for nanowires with straight sidewalls and diameter around 245 nm are 250 kW/cm².

5. Dopant Incorporation

Electrical injection of charge carriers for optical emission requires both efficient formation of free carriers of both doping types and the ability to make ohmic contacts to both types. We have shown that Si doping for growth of *n*-type GaN nanowires is effective, and that Ti:Al contacts reliably produce ohmic behavior for single nanowires contacted on both ends. Typical resistances of single *n*-type nanowires range from a few kΩ to several hundred MΩ. Because it has been quite difficult to independently determine the nanowire surface depletion volume, and the associated surface band bending, mobility, and carrier density, the uncertainties associated with each of these properties are large. Depending on the measurement method applied, apparent carrier concentrations may vary widely. This measurement problem is a topic of ongoing research in many groups, including our own. [Blanchard et al., 2008; Mansfield et al., 2009; Sanford et al., 2010; Stoica and Calarco, 2011]

While growth and contacts for *n*-type nanowires is now routine, a more serious impediment to the fabrication of electrically injected light emitters from GaN nanowires is the achievement of *p*-type doping. The *p*-type problem is compounded by XPS observations of high surface banding for *c*-plane planar *p*-doped GaN which, by extension, suggest that depletion of *p*-type GaN nanowires may be significantly greater than for *n*-type nanowires. [Tracy et al., 2003] Nanowire growth with MBE occurs at high temperatures, where the Mg vapor pressure is high, raising questions as to whether or not the Mg impinging on the surface becomes incorporated into the crystal. The incorporation of Mg in planar MBE growth is known to depend strongly on the V:III ratio, leading to narrow process windows for achievement of reproducible results. The low activation of Mg acceptors at room temperature also requires a relatively high concentration of Mg atoms to achieve even modest free hole concentrations — an atomic density around 3×10^{19} cm^{-3} is needed to achieve a free hole concentration of 1×10^{17} cm^{-3}. These considerations combine to produce a challenge for crystal growth in that the Mg flux and final atomic mole fraction are high enough to affect the crystal growth process itself. One such effect is illustrated for nanowires in Fig. 7(a). Our observation is that high Mg flux leads to increasing diameter as growth progresses, producing morphological difficulties in contacting the nanowires and a large fraction of nanowire regions that are likely depleted of free charge throughout their entire diameter. Unlike the case for *n*-type nanowires, ohmic behavior for a single *p*-type nanowire contacted on both ends has yet to be demonstrated, at least for MBE-grown material, although rectifying behavior is routinely observed. The inability to reliably contact Mg-doped nanowires and the low currents present for quasi-rectifying behavior make it difficult to determine the cause of the high resistance. Possible causes include (1) the amount of Mg incorporated may be either too high or too low, (2) the surface may be trapping sufficient positive charge to mostly deplete the nanowires, or (3) the contact metallization may be unsuited to the nanowire morphology.

Using two separate noncontact methods, we have demonstrated that Mg is indeed present in the crystal. Figure 7(b) shows the low-temperature PL spectrum of some typical Mg-doped nanowires. The spectrum contains the acceptor-bound exciton peak (A^0X_A) and the broad UVL peak associated with Mg doping in GaN films. [Monemar et al., 2010] We have also examined GaN nanowire samples with Secondary Ion Mass Spectroscopy (SIMS) by use of a commercial analysis service. The range of Mg atomic concentrations measured is from around 10^{19} cm^{-3} to 10^{21} cm^{-3}, covering the range observed to generate free holes in planar films. For these data, the Mg ion concentrations were normalized to the Ga ion concentrations, and this ratio was interpreted with calibration data for Mg-doped planar GaN standards. The nanowires did not etch evenly during SIMS analysis, and melting and decomposition were apparent in the SIMS crater [Fig. 8(b)], so the absolute value of the Mg concentration and its depth distribution cannot be considered as reliable as the results obtained in planar films.

In the course of our investigations, we have identified additional impediments to ohmic contact formation on Mg-doped nanowires that are independent of the nanowire free-hole concentration — the three-dimensional nature of the nanowire, the metallurgy of the contacts themselves, and the poor adhesion of metals to insulating layers (such as SiO_x and SiN_x) that are needed to contact dispersed nanowires. The most common metal contacts made to planar *p*-type GaN are thin NiAu contacts, typically 10 nm of each metal, which are annealed in air or in N_2:O_2 mixtures. To completely cover the nanowire, the thicknesses of these layers must be increased by almost an order of magnitude, to 50 nm of Ni and 150 nm of Au. The increased thickness degrades the contact resistance

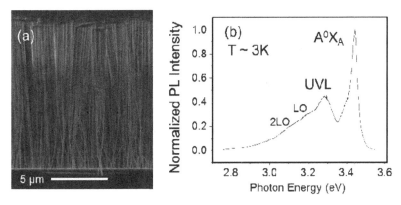

Fig. 7. (a) SEM micrograph of Mg-doped GaN nanowires, showing the high degree of tapering during growth, producing thin roots and thick, coalesced tips. (b) Typical PL spectrum for Mg-doped GaN nanowires. Spectroscopic feature designations are the same as the usage found in [Reshchikov and Morkoç, 2005].

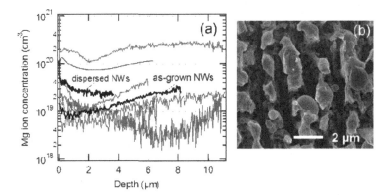

Fig. 8. SIMS analysis of Mg-doped GaN nanowires. (a) Depth profiles for several nanowire runs, with one run (heavy black lines) examined both as-grown and as a thick clump of dispersed nanowires. (b) SEM micrograph of SIMS crater illustrating uneven etching and partial melting of the sample.

in control experiments on planar *p*-type GaN. The large amount of chemical reaction and metal interdiffusion that occur during contact anneals is important for optimizing contact resistance, but also leads to rough, fragile contacts. We have also observed that the nanowires behave as a nucleation site for large void formation during anneals, as shown in Fig. 9. The contact in the immediate vicinity of the nanowire is therefore mechanically weak and much smaller in area than expected. The NiAu contacts also adhere poorly to the underlying SiN_x insulation layer, a phenomenon that enables the study of the underside of the contacts but also makes for contacts that easily scratch and peel. Our attempts to use indium as an alternative metallization scheme met with similar difficulties; in metal contacts condense into isolated islands when heated on the SiO_x insulating layer, and the addition of wetting-layer materials or Au improved the morphology but would not produce low-resistivity contacts with planar *p*-GaN films.

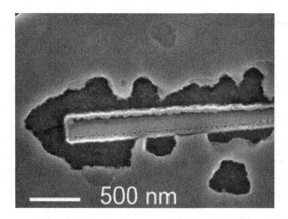

Fig. 9. Underside of NiAu contact to GaN nanowire observed after removal of the metal film and nanowire from the dispersal substrate, illustrating void formation and drastic reduction in contact area with the nanowire.

Despite the large number of remaining engineering issues, we have succeeded in making single nanowire axial *p-n* devices that produce electroluminescence that originates from the metallurgical junction and with peak intensities near the band gap of GaN. Although the luminescence is easily observed even from a single nanowire, the power conversion efficiency of these devices is generally less than 0.01%. However, we also routinely see electroluminescence with comparable low efficiency arising from minority-carrier injection at a Schottky barrier contact to a Mg-doped nanowire. This electroluminescence is completely independent of the presence of a *p-n* junction, and has been observed in GaN devices since the early days of LED development. [Maruska and Stevenson, 1974] Achievement of higher-efficiency, single-nanowire LEDs will require both improved *p*-type doping processes and improved contacts, and will undoubtedly follow an evolutionary path similar to that of the now comparatively mature planar GaN LED technology.

6. Summary

GaN nanowires offer great promise for nanoscale light emitters. Their extremely low defect density, natural optical-cavity formation, and morphology offer applications that cannot be duplicated via conventional methods. Significant progress has been made in diameter control for diameters of 200 nm or less by use of selective epitaxy on patterned masks. The growth of n-type material with low resistivity is now straightforward, despite the fact that methods for determination of absolute values for carrier concentration, mobility, and surface band bending continue to have large uncertainties. Recent observations of UV light emission from electrically injected single nanowires, even though weak in terms of absolute energy conversion efficiency, confirm that there are no inherent limitations to fabrication of LEDs from these materials. The greatest gains are likely to come from improvements in p-type doping and contact metallization, with improved surface passivation also playing an important part.

References

Amano, H., Kito, M., Hiramatsu, K. and Akasaki, I. (1989). P-Type Conduction in Mg-Doped GaN Treated with Low-Energy Electron Beam Irradiation (LEEBI), *Jpn. J. Appl. Phys. Part 2* 28, L2112.

Bertness, K. A., Roshko, A., Mansfield, L. M., Harvey, T. E. and Sanford, N. A. (2007). Nucleation conditions for catalyst-free GaN nanowires, *J. Cryst. Growth* 300, 94-99.

Bertness, K. A., Roshko, A., Mansfield, L. M., Harvey, T. E. and Sanford, N. A. (2008). Mechanism for spontaneous growth of GaN nanowires with molecular beam epitaxy, *J. Cryst. Growth* 310, 3154-3158.

Bertness, K. A., Sanders, A. W., Rourke, D. M., Harvey, T. E., Roshko, A., Schlager, J. B. and Sanford, N. A. (2010). Controlled Nucleation of GaN Nanowires Grown with Molecular Beam Epitaxy, *Advanced Functional Materials* 20, 2911-2915.

Bertness, K. A., Sanford, N. A. and Davydov, A. V. (2010). GaN Nanowires Grown by Molecular Beam Epitaxy, *IEEE Journal of Selected Topics in Quantum Electronics* 17(4), 847-858.

Blanchard, P. T., Bertness, K. A., Harvey, T. E., Mansfield, L. M., Sanders, A. W. and Sanford, N. A. (2008). MESFETs Made From Individual GaN Nanowires, *IEEE Trans. Nanotechnol.* 7(6), 760-765.

Brubaker, M. D., Levin, I., Davydov, A. V., Rourke, D. M., Sanford, N. A., Bright, V. M. and Bertness, K. A. (2011). Effect of AlN buffer layer properties on the morphology and polarity of GaN nanowires grown by molecular beam epitaxy, *J. Appl. Phys.* 110, 053506.

Burnham, S. D., Namkoong, G., Henderson, W. and Doolittle, W. A. (2005). Mg doped GaN using a valved, thermally energetic source: enhanced incorporation, and control, *J. Cryst. Growth* 279(1-2), 26-30.

Calleja, E., Ristic, J., Fernandez-Garrido, S., Ceruffi, L., Sanchez-Garcia, M. A., Grandal, J., Trampert, A., Jahn, U., Sanchez, G., Griol, A. and Sanchez, B. (2007). Growth, morphology, and structural properties of group-III-nitride nanocolumns and nanodisks, *Physica Status Solidi B-Basic Solid State Physics* 244(8), 2816-2837.

Consonni, V., Knelangen, M., Geelhaar, L., Trampert, A. and Riechert, H. (2010). Nucleation mechanisms of epitaxial GaN nanowires: Origin of their self-induced formation and initial radius, *Phys. Rev. B* 81(8), 085310.

Detchprohm, T., Zhu, M. W., Li, Y. F., Zhao, L., You, S., Wetzel, C., Preble, E. A., Paskova, T. and Hanser, D. (2010). Wavelength-stable cyan and green light emitting diodes on nonpolar m-plane GaN bulk substrates, *Appl. Phys. Lett.* 96(5).

Huang, S. J., Xian, Y. L., Fan, B. F., Zheng, Z. Y., Chen, Z. M., Jia, W. Q., Jiang, H. and Wang, G. (2011). Contrary luminescence behaviors of InGaN/GaN light emitting diodes caused by carrier tunneling leakage, *J. Appl. Phys.* 110(6).

Kishino, K., Sekiguchia, H. and Kikuchi, A. (2009). Improved Ti-mask selective-area growth (SAG) by rf-plasma-assisted molecular beam epitaxy demonstrating extremely uniform GaN nanocolumn arrays, *J. Cryst. Growth* 311(7), 2063-2068.

Li, Y. F., You, S., Zhu, M. W., Zhao, L., Hou, W. T., Detchprohm, T., Taniguchi, Y., Tamura, N., Tanaka, S. and Wetzel, C. (2011). Defect-reduced green GaInN/GaN light-emitting diode on nanopatterned sapphire, *Appl. Phys. Lett.* 98(15).

Mansfield, L. M., Bertness, K. A., Blanchard, P. T., Harvey, T. E., Sanders, A. W. and Sanford, N. A. (2009). GaN Nanowire Carrier Concentration Calculated from Light and Dark Resistance Measurements, *J. Electron. Mat.* 38(4), 495-504.

Maruska, H. P. and Stevenson, D. A. (1974). Mechanism of light production in metal-insulator-semiconductor diodes; GaN:Mg violet Light-emitting diodes, *Solid State Electronics* 17, 1171-1179.

Monemar, B., Paskov, P. P., Pozina, G., Hemmingsson, C., Bergman, J. P., Amano, H., Akasaki, I., Figge, S., Hommel, D., Paskova, T. and Usui, A. (2010). Mg-related acceptors in GaN, *Phys. Stat. Sol. (C)* 7(7-8), 1850-1852.

Nakamura, S., Mukai, T., Senoh, M. and Iwasa, N. (1992). Thermal Annealing Effects on P-Type Mg-Doped GaN Films, *Jpn. J. Appl. Phys. Part 2* 31, L139.

Reshchikov, M. A. and Morkoç, H. (2005). Luminescence properties of defects in GaN, *J. Appl. Phys.* 97, 061301.

Ristić, J., Calleja, E., Fernandez-Garrido, S., Cerutti, L., Trampert, A., Jahn, U. and Ploog, K. H. (2008). On the mechanisms of spontaneous growth of III-nitride nanocolumns by plasma-assisted molecular beam epitaxy, *J. Cryst. Growth* 310(18), 4035-4045.

Sánchez-García, M. A., Calleja, E., Monroy, E., Sanchez, F. J., Calle, F., Munoz, E. and Beresford, R. (1998). The effect of the III/V ratio and substrate temperature on the morphology and properties of GaN- and AlN-layers grown by molecular beam epitaxy on Si(111), *J. Cryst. Growth* 183(1-2), 23-30.

Sanford, N. A., Blanchard, P. T., Bertness, K. A., Mansfield, L., Schlager, J. B., Sanders, A. W., Roshko, A., Burton, B. B. and George, S. M. (2010). Steady-state and transient photoconductivity in c-axis GaN nanowires grown by nitrogen-plasma-assisted molecular beam epitaxy, *J. Appl. Phys.* 107(3), 034318.

Schlager, J. B., Bertness, K. A., Blanchard, P. T., Robins, L. H., Roshko, A. and Sanford, N. A. (2008). Steady-state and time-resolved photoluminescence from relaxed and strained GaN nanowires grown by catalyst-free molecular-beam epitaxy, *J. Appl. Phys.* 103(12), 124309.

Schlager, J. B., Sanford, N. A., Bertness, K. A., Barker, J. M., Roshko, A. and Blanchard, P. T. (2006). Polarization-resolved photoluminescence study of individual GaN nanowires grown by catalyst-free MBE, *Appl. Phys. Lett.* 88, 213106.

Schlager, J. B., Sanford, N. A., Bertness, K. A. and Roshko, A. (2011). Injection-level-dependent internal quantum efficiency and lasing in low-defect GaN nanowires, *J. Appl. Phys.* 109(4), 044312.

Stoica, T. and Calarco, R. (2011). Doping of III-Nitride nanowires grown by molecular beam epitaxy, *IEEE J. Sel. Topics Quantum Electron.* nn.

Tracy, K. M., Mecouch, W. J., Davis, R. F. and Nemanich, R. J. (2003). Preparation and characterization of atomically clean, stoichiometric surfaces of n- and p-type GaN(0001), *J. Appl. Phys.* 94(5), 3163-3172.

van Vugt, L. K., Ruhle, S. and Vanmaeklbergh, D. (2006). Phase-correlated nondirectional laser emission from the end facets of a ZnO nanowire, *Nano Lett.* 6(12), 2707-2711.

VanMil, B. L., Guo, H., Holbert, L. J., Lee, K., Swartz, C. H., Liu, T., Korakakis, D. and Myers, T. H. (2005). High temperature limitations for GaN growth by RF-plasma assisted molecular beam epitaxy: Effects of active nitrogen species, surface polarity, and excess Ga-overpressure, *Phys. Stat. Sol. (C)* 2(7), 2174-2177.

Yoshizawa, M., Kikuchi, A., Mori, M., Fujita, N. and Kishino, K. (1997). Growth of self-organized GaN nanostructures on Al2O3(0001) by RF-radical source molecular beam epitaxy, *Jpn. J. Appl. Phys. Part 2* 36(4B), L459-L462.

Zhang, Y. Y. and Yao, G. R. (2011). Performance enhancement of blue light-emitting diodes with AlGaN barriers and a special designed electron-blocking layer, *J. Appl. Phys.* 110(9).

RECESSED-GATE NORMALLY-OFF GaN MOSFET TECHNOLOGIES

KI-SIK IM[*], KI-WON KIM[*], DONG-SEOK KIM[*], HEE-SUNG KANG[*], DO-KYWN KIM[*,†],
SUNG-JAE CHANG[†], YOUNG-HO BAE[‡], SUNG-HO HAHM[*],
SORIN CRISTOLOVEANU[†] and JUNG-HEE LEE[*]

[*]*School of Electrical Engineering and Computer Science, Kyungpook National University, Daegu, Korea*
[†]*IMEP-LAHC, Grenoble Institute of Technology, Minatec, BP 257, 38016 Grenoble Cedex 1, France*
[‡]*Department of Electronic Engineering, Uiduk University, Gyeongju, Korea*
ksim@ee.knu.ac.kr; kwbest@ee.knu.ac.kr; gdolgun@ee.knu.ac.kr; hskang@ee.knu.ac.kr;
kdky1@nate.com; changes@minatec.inpg.fr; yhbae@uu.ac.kr; shhahm@ee.knu.ac.kr;
sorin@enserg.fr; jlee@ee.knu.ac.kr

We have fabricated and investigated several types of GaN MOSFETs with normally-off operation. The recessed-gate GaN MOSFET is preferred for normally-off operation, because the threshold voltage (V_{th}) of the device can be easily controlled, but it suffers from relatively modest current drivability which must be improved by adopting appropriate device structure and/or process. Enhanced performances have been achieved in this work by combining the recessed-gate technology with additional processes, such as: the post-recess tetramethylammonium hydroxide (TMAH) treatment to remove the plasma damage, the post-deposition annealing of gate oxide to decrease the gate leakage current, the re-growth of n+ GaN layer for source/drain to improve the access resistance and V_{th} uniformity, the stress control technology to achieve extremely high 2-D electron-gas density (2DEG) on source/drain and decrease the series resistance, and the use of the p-GaN back-barrier to decrease the buffer leakage current. The GaN-based FinFET with very narrow fin was also investigated as a possible candidate for high performance normally-off GaN MOSFETs.

Keywords: Normally-off; GaN; MOSFET; MISFET; HFET; 2DEG; MOCVD; Buffer; TMAH; Al$_2$O$_3$; FinFET.

1. Introduction

The AlGaN/GaN heterojunction field-effect-transistors (HFETs) are very promising for high power and radio frequency applications due to their superior material properties such as high breakdown fields and polarization-induced two-dimensional electron gas (2DEG, formed at AlGaN/GaN heterointerface) with large density and high saturated electron velocity. However, the large 2DEG density makes the AlGaN/GaN HFETs very difficult to be used in normally-off operation. In power switching application, normally-off device operation is essential to simplify the design of driving circuits and to reduce power loss during switching. Many efforts have been dedicated to achieve a successful normally-off operation by investigating various device structures or processes, such as p-GaN/AlGaN/GaN HFET[1~2], tunnel junction HFET[3], and recessed-gate AlGaN/GaN HFET[4]. In addition to the normally-off operation, it is necessary that the power switching devices exhibit high threshold voltage (usually $V_{th} > 3$ V), large gate voltage swing, and low leakage current to improve the switching performances. The GaN MOSFETs, which can be fabricated by depositing a gate insulator on the GaN surface after fully removing the AlGaN layer from the AlGaN/GaN heterostructure in the gate region, are very promising because they easily satisfy most of the requirements mentioned above for power switching applications, even though there are still critical issues to be solved such as relatively low current drivability (mainly caused by the plasma damage in the gate region), uniformity in threshold voltage, and charge trapping/detrapping in the gate insulator.

In this work, we report the effects of a simple tetramethylammonium hydroxide (TMAH) treatment as a post gate-recess process on enhancing the device performance of the GaN MOSFET, and the effects of post-deposition annealing of the Al_2O_3 dielectric on mitigating the charge trapping/detrapping in the gate insulator. We also introduce the GaN MOSFETs with re-grown n+ GaN source/drain which may provide a solution against high access resistance and V_{th} variability. The use of extremely high 2DEG density as source and drain is very effective in obtaining high current drivability and reduced series resistance of the device. Next, the GaN MOSFET with the p-GaN back-barrier is described as a method of obtaining higher V_{th} and lower buffer leakage current. Finally, the fabrication of the GaN-based FinFET is introduced as a very promising normally-off GaN-based MOSFET technology.

2. Normally-off GaN MOSFET with Post-Recess TMAH-Treatment

The schematic cross section of the investigated normally-off GaN MOSFETs is shown in Fig. 1. To fabricate the MOSFETs, the device active region was isolated by transformer-coupled-plasma reactive ion etching (TCP-RIE) with a depth of 250 nm using a BCl_3/Cl_2 gas mixture. The gate region of length of 2.5 μm was then fully recessed with a photoresist mask by completely removing the 25 nm-thick AlGaN layer using the same gas mixture and tetramethylammonium hydroxide (TMAH with 25% concentration) treatment at 85°C for 10 min without removing the photoresist. A 38 nm-thick Al_2O_3 layer as gate insulator was deposited on the TMAH-treated GaN surface by atomic layer deposition (ALD). Ti/Al/Ni/Au ohmic contacts were formed as source and drain electrodes and annealed using rapid thermal process (RTP) at 850°C for 30 sec in N_2 ambient. Finally, a Ni/Au gate electrode was deposited on the gate insulator. The gate length L_g and the gate width W_g, defined by the rectangular recess pattern, were 2.5 and 50 μm, respectively.

The GaN surface exposed after the gate recess showed increased roughness and many sharp protrusions and pits were found on the surface. After the TMAH treatment, however, the recessed GaN surface exhibited less sharpened protrusions and reduced pit numbers, which resulted in improved roughness. The superior surface morphology is because the TMAH solution effectively etches the side slopes of the sharp etch protrusions, then the etch slows down due to the anisotropic etching property of the TMAH solution as the slopes become blunt, and eventually the etching hardly proceeds once some of the slopes are removed and the surface becomes more or less smooth. At same time, this surface etching with TMAH treatment can help to eliminate the surface damage caused by plasma etching which improves the interface quality between the Al_2O_3 gate insulator and the GaN surface.

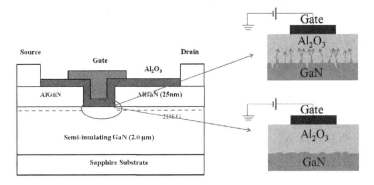

Fig. 1. Schematic cross section of the GaN MOSFET; the arrow lines show the models of Al_2O_3/GaN interface at forward biases with (lower) and without (upper) TMAH treatment.

The gate leakage characteristics of the fabricated GaN MOSFETs are shown in Fig. 2 (a). It is noticed that the difference in leakage current between the devices with and without TMAH treatment is remarkable especially in the forward bias region. The leakage current of the TMAH-treated device was measured as 10^{-9} A/mm at $V_g = 15$ V, which is lower by approximately 6 orders in magnitude compared to the value for the device without TMAH treatment. The reason for this large improvement in the forward gate leakage currents between two devices can be explained as follows: The sharp protrusions formed on the GaN surface after the gate recess increase the field enhancement factor (or decrease the barrier height) of the GaN surface and enhance the electron emission from the surface. This is effective only at forward biases for the devices without TMAH treatment, as modeled in the inset of Fig. 1. Instead, the GaN surface becomes smooth after the TMAH treatment so that electron emission is inhibited due to a decreased field enhancement factor at the surface. This explains why the TMAH-treated devices exhibit very small leakage current characteristics regardless of the bias conditions. It is, therefore, very important to keep the roughened GaN surface smooth with an appropriate post recess process such as TMAH treatment.

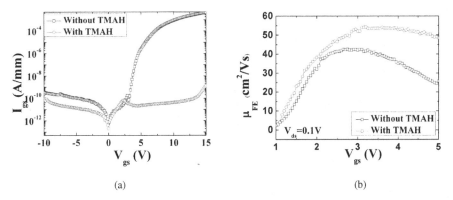

(a)

(b)

Fig. 2. (a) Gate leakage current and (b) field-effect mobility versus gate bias for GaN MOSFETs fabricated with and without TMAH treatment.

The maximum field-effect mobility (μ_{FE}) of the TMAH-treated device, extracted in the linear region (at $V_d = 0.1$ V) as shown in Fig. 2 (b), was increased to 55 cm$^2 \cdot$V$^{-1} \cdot$s^{-1} from 42 cm$^2 \cdot$V$^{-1} \cdot$s^{-1} for the device without TMAH treatment. This clearly indicates that the improved GaN surface morphology after the TMAH treatment leads to less electron scattering at the interface and hence higher electron mobility in the channel.

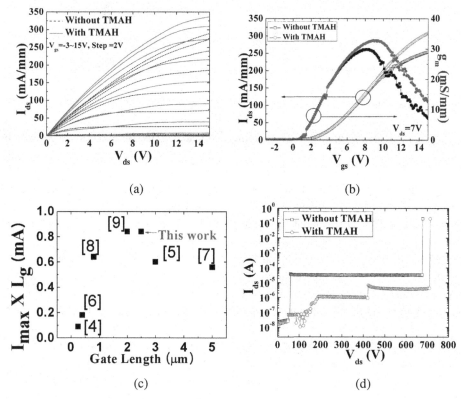

Fig. 3. (a) Output I_d-V_d characteristics, (b) transfer curves, (c) the $I_{dmax} \times L_g$ product versus gate length, and (d) breakdown voltages of the fabricated devices.

The characteristics of the fabricated GaN MOSFETs with $L_g = 2.5$ µm and $L_{gd} = 35$ µm are shown in Fig. 3 (a) and (b). Both devices successfully demonstrated normally-off operation with V_{th} of 3.5 V, defined as the gate bias intercept of the linear extrapolation of drain current at the point of peak transconductance (g_m). The maximum drain current ($I_{d,max}$) at a gate voltage of 15 V and the maximum g_m for TMAH-treated device were increased to 336 mA/mm and 33 mS/mm from the corresponding values of 289 mA/mm and 30 mS/mm for the device without TMAH treatment. These gains in $I_{d,max}$ and g_m are definitely due to the improved channel mobility caused by the smooth surface with TMAH treatment. The $I_{dmax} \times L_g$ product of 0.84 for the TMAH-treated device, measured at $V_d = V_g = 15$ V, is among one of the highest values ever reported for the normally-off GaN MOSFETs as shown in the Fig. 3 (c)[4-9]. The calculated $R_{ON,sp}$ was also decreased from 14.3 to 10.9 m$\Omega\cdot$cm^2 with TMAH treatment. Figure 3 (d) shows the off-state (V_g = 0 V) characteristics of the fabricated devices. The breakdown voltage was defined as the drain voltage at which the drain current reaches 1 mA/mm in off-state. The TMAH-treated device exhibited higher off-state breakdown voltage of 750 V in contrast to 650 V for the device without TMAH treatment. It is not clear at this point why the breakdown voltage of the device with TMAH treatment is higher. A possible explanation is that the field strength at the drain-end of the gate, which generally determines the breakdown voltage of the device, becomes weaker when the rough GaN surface is smoothed with TMAH treatment.

3. Post-Deposition Annealing Effect in Normally-off GaN MOSFET

Al$_2$O$_3$ deposited by ALD is a promising gate dielectric material[10]. Al$_2$O$_3$ has a large band gap (9 eV), a high dielectric constant (8.6~10), and a high breakdown field (10 MV/cm)[11]. Additionally, ALD is a layer-by-layer controlled process offering a precise deposition at atomic scale and a good film uniformity. In this section, we present a normally-off GaN MOSFET featuring plasma-assisted atomic-layer-deposited (PAALD) Al$_2$O$_3$ gate oxide on the fully recessed surface of the gate region. We investigate the effects of post-deposition annealing of the Al$_2$O$_3$ dielectric on the charge trapping/detrapping in the gate insulator.

A schematic cross-section of the fabricated normally-off GaN MOSFET is shown in Fig. 4. For the characterization of the oxide/semiconductor interface, capacitance-voltage (C-V) measurements of the virtual gate capacitors, fabricated on n-GaN (n = 1 $\times$ 10^{16} cm^{-3}), were performed. The samples were patterned with circular capacitors with a diameter of 40 μm and dual metal layers of Ni/Au (40/50 nm) were deposited on both gate and virtual ground contact. Virtual ground contact area is much larger than the actual size of the measured capacitance. In this case the series capacitance from the virtual ground can be neglected during measurement and only the small capacitor defined by the gate area remains. This method is not only a great advantage in fabrication, because no second metallization step is needed, but also guarantees that no parasitic capacitances of ohmic contacts are included.

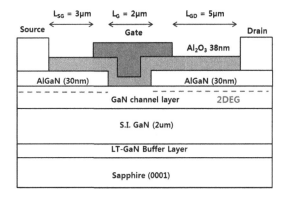

Fig. 4. Schematic cross-section of the proposed Al$_2$O$_3$/GaN MOSFETs.

Aluminum oxide has to withstand the thermal budget of subsequent process steps. It is known that around 850°C the amorphous film becomes polycrystalline. Even though the dielectric constant of crystalline films is higher (around 12 compared to 8 for amorphous films), the leakage current increases by a few orders of magnitude, which makes the dielectric useless for thin film applications. From XRD measurements, shown in Fig. 5, we could find a small peak, indicating long range order of polycrystal, after 20 min annealing at 850°C. Figure 6 shows the gate leakage currents of the fabricated Al_2O_3/GaN MOSFETs. The measured leakage currents were below a few tenth of pA/mm up to $V_g = 10$ V and did not show any significant difference with the various annealing temperatures. However, the 850°C annealed device shows breakdown at a gate voltage of 8 V. This result coincides with the XRD phase product.

We conclude that 800°C is the maximum annealing temperature for 35 nm thick Al_2O_3 film and further increase in the anneal temperature results in an abrupt phase change of the Al_2O_3 film from amorphous to polycrystalline and hence, a rapid increase in leakage current. This result also matches with the data from other reports[12].

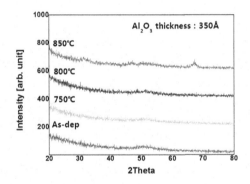

Fig. 5. Results of XRD phase analysis of Al_2O_3 dielectric for various annealing temperature.

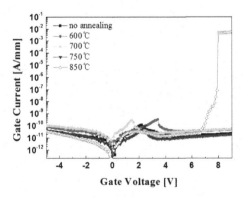

Fig. 6. Gate current versus gate voltage in the fabricated Al_2O_3/GaN MOSFETs.

   The typical C-V characteristics measured by using constant-voltage stress for 1 min at 3 V and room temperature are shown in Fig. 7 for non-annealed Al_2O_3 sample (triangle) and annealed at 750°C (square), respectively. The solid line is the ideal curve. The shifts of the C-V curves from the ideal one, -0.11 V for non-annealed sample and -0.42 V for post-deposition annealed sample were caused by the fixed charge in the oxide film. It is assumed that after annealing, oxygen deficiency in Al_2O_3 film near the interface, caused by outdiffusion of oxygen into the GaN or air, leads to net positive fixed charge in the oxide film.

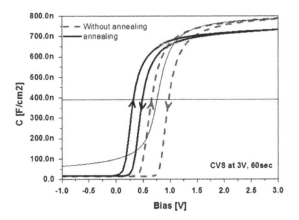

Fig. 7. C-V characteristics of the virtual gate capacitors measured at 100 KHz.

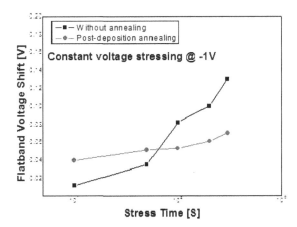

Fig. 8. Dependence of flatband voltage shift (ΔV_{FB}) as a function of stress time.

   The hysteresis of the post-deposition annealed sample was lesser than that of the non-annealed sample. The clock-wise hysteresis induced by positive gate bias (negative shift of curve) implies electron injection into traps within the Al_2O_3 oxide. The flat band voltage shift with increasing stress time can be neglected in post-deposition annealed samples as shown in Fig. 8. This indicates that post-deposition annealing of Al_2O_3 can effectively reduce the trap density which results in improved device reliability. In addition, according to our previous experiment[13], the interface trap density was high throughout the whole band-gap energy for the non-annealed sample. But the trap density near the mid-gap was effectively decreased after annealing the sample at a high temperature above 750°C. Also, the interface trap density of Al_2O_3/GaN is rather low compared to that of other insulators.

The fabricated GaN MOSFETs demonstrate good normally-off operation with V_{th} of about 1.4 V, independent of post-deposition annealing. The V_{th} was extrapolated from the transfer current-voltage (I-V) characteristics in the linear region (at $V_d = 0.1$ V) as shown in Fig. 9 (a). The μ_{FE} extracted from the g_m maximum was 50 $cm^2 \cdot V^{-1} \cdot s^{-1}$ before annealing and increased to 55 $cm^2 \cdot V^{-1} \cdot s^{-1}$ after annealing. This increment of μ_{FE} is due to the decrease in the trap density at the Al_2O_3/GaN interface after post-deposition annealing, leading to reduced interface scattering of the channel electrons. The subthreshold slope, which impacts the switching capability, is also closely related to the interface trap capacitance (C_{it}). The swing decreased from 640 mV/dec (without post-deposition annealing) to 350 mV/dec after annealing at 750°C. The interface state density (D_{it}), estimated from the subthreshold slope, was reduced from 1.4×10^{13} to 6.7×10^{12} cm^{-2} after annealing at 750°C. The magnitude of off-state current was 6 µA/mm before annealing and two orders of magnitude lower (0.1 µA/mm) after annealing. Figure 9 (b) shows I_d-V_d characteristics of the GaN MOSFETs. The saturation drain current (330 mA/mm at $V_g = 15$ V) and the maximum g_m (32 mS/mm at $V_d = 10$ V) do not reveal significant variations after anneal.

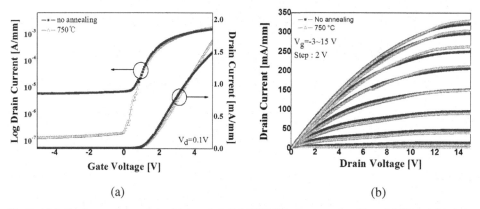

(a)   (b)

Fig. 9. (a) I_d-V_g characteristics of the fabricated GaN MOSFET in linear region ($V_d = 0.1$ V) and (b) drain current characteristics as a function of drain voltage and gate voltage in GaN MOSFETs before and after anneal.

Figure 10 shows the drain current hysteresis (inset) and the related V_{th} shift as a function of the post deposition annealing temperature of Al_2O_3. A quite large V_{th} shift ($\Delta V_{th} = 0.2 \sim 0.3$ V) was observed according to the direction of the gate voltage sweep. The calculated oxide trap density in the Al_2O_3 was $3 \sim 4 \times 10^{11}$ cm^{-2} before annealing. As the annealing temperature was increased, the ΔV_{th} gradually reduced to finally reach a negligible value after annealing at 750°C. It is concluded that the annealing process is very effective in reducing both the interface state and the oxide trap density in the gate dielectric and enhancing the mobility of the Al_2O_3/GaN MOSFETs.

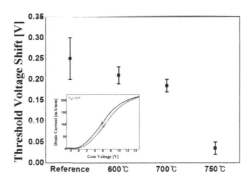

Fig. 10. Hysteresis-related V_{th} shift versus annealing temperature of Al_2O_3.

4. Normally-off GaN MOSFETs with Re-grown Source/Drain

This section introduces two types of GaN MOSFET with raised source and drain (S/D), fabricated using n^+ GaN re-growth with and without prior S/D etching (Fig. 11). An epitaxial 2 μm-thick crack-free and semi-insulating GaN layer was grown on sapphire (0001) substrate by MOCVD at 1070°C. SiO_2 layer with 100 nm was deposited on GaN layer by PECVD as a hard mask (Fig. 11 (a)) before selective dry etching (Fig. 11 (b)) and n^+-GaN re-growth in S/D region (Fig. 11 (c)). MOSFET A was etched down 30 nm by TCP-RIE with BCl_3/Cl_2 gas mixture, and MOSFET B was left unetched. 300 nm-thick Si-doped GaN layers with 5×10^{18} cm^{-3} were selectively grown on the S/D regions of both MOSFETs. The SiO_2 layer was etched and 30-nm-thick Al_2O_3 gate dielectric was deposited by ALD. After Al_2O_3 etching in S/D region, metal layers (Ti/Al/Ni/Au) were deposited by e-beam evaporator and annealed at 900°C for 30 sec in N_2 ambient. Finally, Ni/Au was e-beam evaporated to form the gate ($L_g = 10$ μm and $W_g = 100$ μm).

Figure 12 shows drain current and g_m characteristics for MOSFETs A and B. In the case of MOSFET A, the curves are well behaved and V_{th} is reasonable (≈ 2 V). By contrast, V_{th} of MOSFET B is around 6 V. The MOSFET improvement by S/D etching and regrowth (Process A) is visible in terms of lower V_{th}, higher ON-current, increased g_m and better subthreshold behaviour.

In order to extract accurate device parameters, we used the Y-function, $Y = I_d/\sqrt{g_m}$, method[14]. The low-field mobility in MOSFET A is 32 $cm^2 \cdot V^{-1} \cdot s^{-1}$, almost twice the value in MOSFET B (19 $cm^2 \cdot V^{-1} \cdot s^{-1}$). These results are in reasonable agreement with the μ_{FE} (MOSFET A: 26 $cm^2 \cdot V^{-1} \cdot s^{-1}$ and B: 14 $cm^2 \cdot V^{-1} \cdot s^{-1}$) extracted from the g_m peak, which is underestimated due to series resistance effect. The mobility degradation factors, 0.04 V^{-1} and 0.03 V^{-1} for MOSFET A and B, enabled us to estimate the series resistance R_{SD} (MOSFET A: 6 MΩ and B: 8 MΩ). The selective etching and re-growth of buried S/D

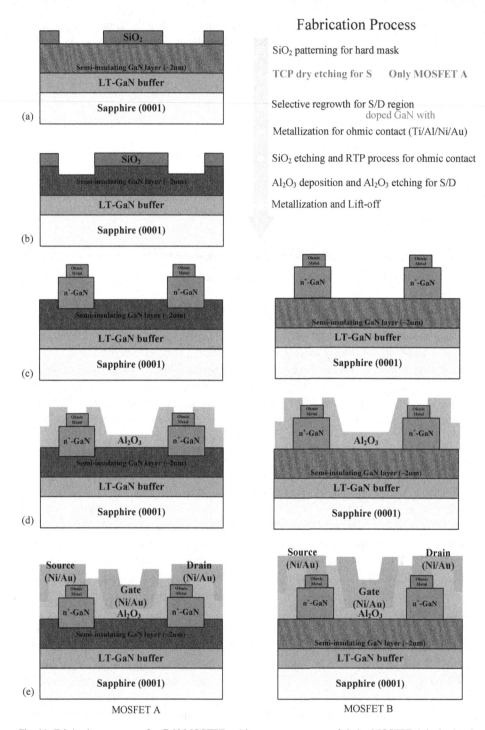

Fabrication Process

SiO$_2$ patterning for hard mask

TCP dry etching for S   Only MOSFET A

Selective regrowth for S/D region
                                        doped GaN with

Metallization for ohmic contact (Ti/Al/Ni/Au)

SiO$_2$ etching and RTP process for ohmic contact

Al$_2$O$_3$ deposition and Al$_2$O$_3$ etching for S/D

Metallization and Lift-off

MOSFET A   MOSFET B

Fig. 11. Fabrication processes for GaN MOSFETs with re-grown source and drain. MOSFET A had prior dry etching while MOSFET B had no etching for S/D regions.

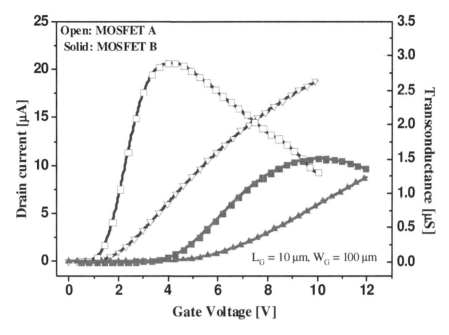

Fig. 12. Drain current and g_m as a function of gate bias in MOSFET A and B (V_D = 50 mV, Hold time = 2 sec, Delay Time = 10 msec, V_G step = 30 mV at 300 K).

region (Process A) result in a reduction in R_{SD}. Note that the carrier path, from the channel to the metal S/D contact, is longer in type B (see Fig. 11 (e)), which also contributes to a higher R_{SD} value.

The subthreshold swing of 0.8 V/dec for MOSFET A is smaller than for MOSFET B which correlates with the difference in leakage current at off-state (one order of magnitude lower in MOSFET A, as shown in Fig. 13).

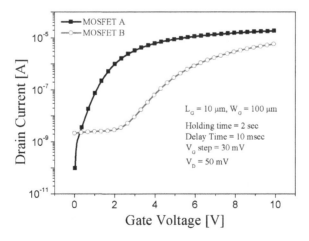

Fig. 13. Log scale drain current versus gate voltage in MOSFET A and B.

Figure 14 shows I_d (V_g) and g_m (V_g) characteristics of MOSFET A from 77 K to 300 K. While V_{th} increases as expected at lower temperature, the drive current and g_m surprisingly decrease. The mobility degradation at low temperature contrasts with the case of SOI or Si MOSFETs, where the mobility is significantly improved ($\mu \propto T^{-1}$). This suggests that the beneficial effect of reduced phonon scattering is masked by a different mechanism. Shallow and deep traps attributed to nitrogen vacancy generation during the MOCVD growth may be involved[15-16]. Similar trends were measured for MOSFET B.

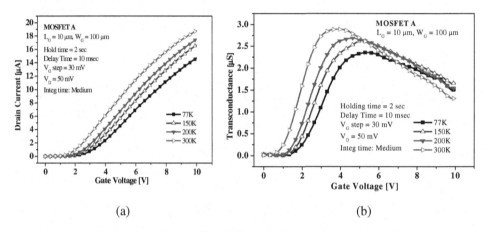

(a)   (b)

Fig. 14. (a) Linear scale drain current and (b) g_m as a function of gate bias in MOSFET A.

Additional measurements indicate that the re-growth technology can provide a solution for obtaining more uniform threshold voltage on the wafer.

5. Normally-off GaN MOSFET with Extremely High 2DEG Density on Source/Drain

In this section, we examine another solution for engineering the source and drain of normally-off GaN MOSFETs. We propose the stress-control technology to achieve higher 2DEG on source/drain, enabling lower series resistance and higher on-current. The 2DEG density (1.8×10^{14} cm^{-2}) in source and drain is at least an order higher than that of normal AlGaN/GaN structure. This extremely high 2DEG density reduces both the parasitic and the on-resistance of the device, which is very important in power switching operation. To obtain a high 2DEG density at the AlGaN/GaN interface we have utilized large tensile stress, induced during the growth of the GaN layer on silicon substrate. Usually, the tensile stress must be minimized in order to guarantee high quality AlGaN/GaN heterostructure. However, the tensile stress can be very useful in increasing the 2DEG density. The larger tensile stress in GaN film enhances piezoelectric effect at AlGaN/GaN interface, which results in great increase in the polarization-induced 2DEG density.

For the device fabrication, AlGaN/GaN heterostructure material with mobility of 120 cm^2/V·s and 2DEG density of 1.8 × 10^{14} /cm^2 was prepared. The MOSFET fabrication was similar to that described in section 2 and the schematic cross section is shown in Fig. 15. The gate length defined by the recess width was 0.25 μm, and the source-drain and gate-source distances were 6.8 and 1.5 μm, respectively.

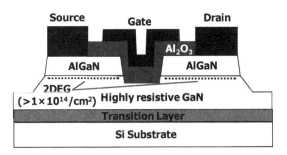

Fig. 15. Schematic cross section of the layer structure and the proposed fully recessed GaN MOSFET. The stress in the GaN layer can be controlled by varying the thickness of the transition layer.

Figure 16 shows the measured DC characteristics of the GaN MOSFET which demonstrate successful normally-off operation. V_{th} extrapolated from I-V curve is 2.0 V in the saturation region. Maximum drain current and extrinsic g_m are as high as 353 mA/mm and 98 mS/mm, respectively. This confirms the benefit of the high 2DEG density in source and drain for enhanced device performance. The calculated maximum μ_{FE} is 225 cm^2/V·s, which is the highest value ever reported for GaN MOSFET. The subthreshold swing is 317 mV/decade in the linear region, comparable to other normally-off GaN devices.[5,7,27] OFF-state breakdown voltage (at V_g = 0 V) is 39 V, which is relatively low due to the very high 2DEG density in the drain. The breakdown voltage can be improved by applying multiple recess or field plate.

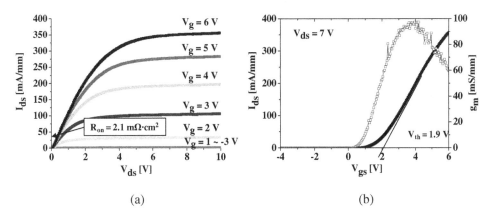

(a)   (b)

Fig. 16. (a) I_d-V_d characteristics and (b) transfer characteristic at V_d = 7 V of the strained GaN MOSFET.

6.   Normally-off AlGaN/GaN MOSFET with p-GaN Back Barrier

The design of a buffer layer for achieving low buffer leakage (off-state leakage current) is very important: the switching performance of the device can be enhanced while reducing the off-state power consumption. An unintentionally doped highly resistive GaN buffer layer or an Fe- or C-doped semi-insulating GaN buffer layer have been used to reduce the leakage current[17-19]. Also, an InGaN or AlGaN back-barrier has been introduced between the channel and the buffer layer to further increase the carrier confinement in the channel[20-21]. Numerical simulations showed the role of p-type GaN barrier in GaN transistors with a configuration different from that in Fig. 17[22]. However, most of these buffer layers and back-barriers still suffer from problems related to the insufficient potential barrier height. The introduction of Mg-doped p-GaN as a back-barrier in the buffer layer of the AlGaN/GaN heterostructure may be an ideal approach to the reduction of the buffer leakage current. The reason is that the p-GaN layer forms an extremely high potential barrier for the channel electrons, preventing them from flowing through the buffer layer[23]. However, the use of the p-GaN back-barrier faces serious problems: diffusion of Mg atoms into the channel layer, memory effect, depletion of the 2DEG channel, and electron mobility degradation in the channel formed at the AlGaN/GaN interface.

In our devices, a high quality p-GaN back-barrier was introduced under a 250-nm-thick undoped GaN channel layer (Fig. 17). The 0.2-μm-thick p-GaN back-barrier exhibited p-type conductivity with a hole concentration of 2.0×10^{17} cm^{-3} and a mobility of 10 cm$^2 \cdot$V$^{-1} \cdot$s^{-1} after in situ activation at 800°C in an MOCVD chamber. The estimated electron mobility and density in 2DEG channel of AlGaN/GaN heterostructure were 275 cm$^2 \cdot$V$^{-1} \cdot$s^{-1} and 5.48×10^{12} cm^{-2}, respectively.

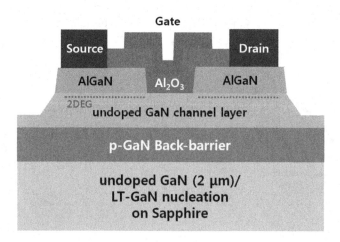

Fig. 17. Cross-sectional schematic of normally-off GaN MOSFET with p-GaN back-barrier.

It is important that the thickness of the undoped GaN channel layer is optimized to achieve a proper normally-off operation and to minimize the memory effect of Mg doping. A depletion layer approximately 250 nm thick is formed when the undoped GaN channel layer is grown on the p-GaN back-barrier[24]. If the thickness of the undoped GaN layer is less than 250 nm, the current level becomes too low because electrons cannot be sufficiently supplied from the source into the channel, due to the depletion of the 2DEG channel. On the other hand, if the thickness exceeds 250 nm, the buffer leakage current will increase, which is undesirable for power-switching applications. Figure 18 shows the secondary ion mass spectroscopy (SIMS) depth profile of Mg atoms. The out-diffusion thickness of Mg atoms into the undoped GaN layer is about 100 nm, indicating that the 2DEG channel at the AlGaN/GaN interface is not severely affected by Mg out-diffusion if the GaN channel thickness is 250 nm.

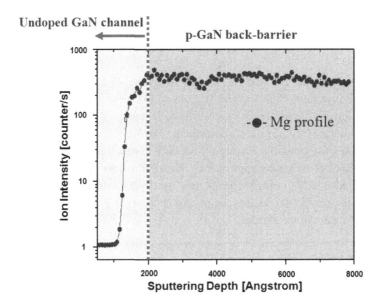

Fig. 18. SIMS depth profile of the Mg atoms for the stack of undoped GaN channel and p-GaN back-barrier.

The buffer leakage current of the designed AlGaN/GaN heterostructure with the p-GaN back-barrier is about 10^3 orders of magnitude lower than that of the conventional AlGaN/GaN heterostructure without the p-GaN back-barrier, as shown in Fig. 19. This improvement was expected because the undoped-GaN channel layer becomes totally depleted, owing to the extremely high potential barrier caused by the p-GaN layer. The inset of Fig. 19 shows the simulated band diagrams of AlGaN/GaN heterostructure with and without p-GaN back-barrier, as obtained using a 1D Poisson solver. The existence of a high-potential p-GaN back-barrier is evident.

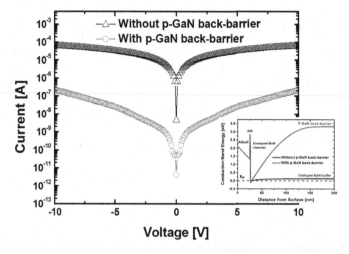

Fig. 19. Buffer leakage current with and without p-GaN back-barrier. The leakage current is significantly reduced with the p-GaN layer acting as a very high back-barrier for the channel electrons (inset).

Figure 20 shows the DC and transfer I-V characteristics in the saturation region at V_d = 6 V. The maximum drain current (I_{dmax}) and extrinsic g_m were 90 mA/mm and 30 mS/mm, respectively. These results successfully demonstrate the normally-off operation with a V_{th} as high as 2.9 V. The high-potential p-GaN back-barrier contributes to increase the V_{th} of AlGaN/GaN-based normally-off MOSFET, which is one of the most important issues in GaN power-switching applications. The preliminary device performance suffers from the degraded 2DEG channel properties by depletion effect and interface quality damage from plasma etching in gate region. An appropriate surface treatment can recover the damage and smooth the surface.

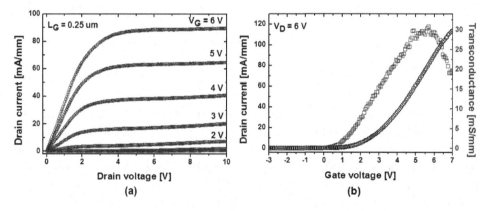

Fig. 20. Experimental (a) DC characteristics and (b) transfer I-V characteristics in the saturation region (at V_d = 6 V) of a GaN MOSFET with p-GaN back barrier. The V_{th} reaches 2.9 V.

7. GaN-based FinFET with Normally-off Operation

The GaN-based FinFET with very narrow fin was also investigated as a possible candidate for high performance normally-off GaN MOSFETs. For GaN-based FinFETs, several research groups have reported methods of achieving positive V_{th} using the AlGaN/GaN-based tri-gate FinFETs with[25] or without[26] gate insulator, and AlN/GaN-based tri-gate FinFET[27]. In this section, we demonstrate a normally-off GaN FinFET with nanometer-size fin width (Fig. 21).

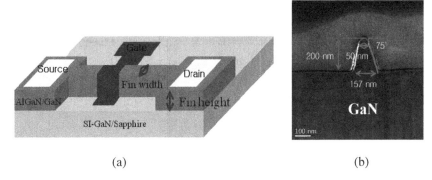

(a)   (b)

Fig. 21. (a) Schematic cross section of the GaN-based FinFET and (b) TEM image.

The GaN FinFET has an AlGaN/GaN heterostructure with high 2DEG density in the top channel. For both device isolation and fin formation, the active region of the device was patterned by electron-beam lithography using hydrogen silsesquioxane (HSQ) and defined by TCP-RIE using a BCl_3/Cl_2 gas mixture. TMAH treatment (25% solution at 85°C) was directly followed to remove the surface roughness from the etched GaN surface[28]. A 20-nm-thick Al_2O_3 gate insulator layer was then formed by ALD. After contact hole opening for the source and drain, Ti/Al/Ni/Au was deposited using an electron-beam evaporator. Finally, Ni/Au was deposited as the gate metal, the gate being 1 μm long. The TEM image of the fabricated GaN-based FinFET with 50 nm fin width is shown in Fig. 21 (b).

Figure 22 shows the DC characteristics of a fabricated GaN FinFET which features 120 nm fin height and 80 nm fin width. The effective channel width is W = (80 + 2 × 120 nm), multiplied by the number of fins in parallel. The GaN FinFET exhibits a I_{dmax} of 1.34 A/mm and a maximum g_m of 662 mS/mm at a gate voltage of 6 V and a drain voltage of 7 V, showing excellent pinch-off characteristics. The on-resistance of this device was estimated to be 0.05 mΩ·cm^2 in the linear region, which is relatively low compared with other results.[4,26,28] These superior device characteristics are mainly related to the strong screening effect due to the high electron concentration inside the narrow fin.

As shown in Fig. 22. (a), this device demonstrates normally-off operation with V_{th} of 4.2 V at $V_d = 7$ V. The reason for normally-off operation is presumably the electron depletion, in the HEMT channel, induced by lateral field from the side gates which results in a positive shift of V_{th}. Besides, the strain relaxation on the edge of nano-sized FinFET weakens the piezoelectric polarization and leads to the reduction of 2DEG density in the channel of FinFET.

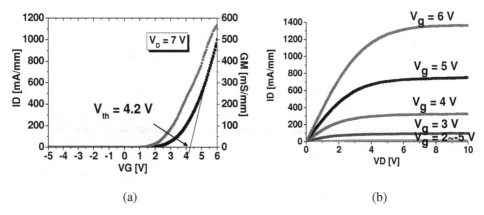

(a)   (b)

Fig. 22. (a) Transfer curves in saturation and (b) I_d-V_d characteristics of fabricated GaN-based FinFET.

The on-off current ratio is 10^6 and the subthreshold swing is 470 mV/dec in the linear region (at $V_d = 0.1$ V). The subthreshold characteristic shown in Fig. 23 (a) clearly reveals a double conduction mechanism. The top channel with AlGaN/GaN heterostructure operates in normally-ON mode and opens at $V_g = -2$ V. For positive gate bias, the lateral MOS channels are gradually enriched whereas the HEMT channel becomes depleted by the lateral gate action. The calculated maximum field-effective mobility was as large as 280 $cm^2 \cdot V^{-1} \cdot s^{-1}$. The gate leakage currents are very low: 10^{-9} A at both $V_g = -5$ V and $V_g = +6$ V, respectively (Fig. 23 (b)).

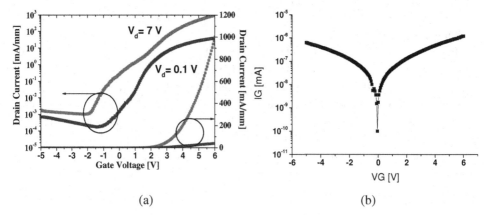

(a)   (b)

Fig. 23. (a) Drain current versus gate voltage in GaN FinFET measured in linear region ($V_d = 0.1$ V) and in saturation region ($V_d = 7$ V). (b) Gate leakage current versus gate voltage.

8. Conclusions

Recessed-gate GaN MOSFETs with successful normally-off operation have been experimentally demonstrated. Several technological modules (Table 1) have been developed and applied to solve practical issues and to further improve the device performance. These variants have been implemented at various stages of maturity of GaN MOSFET technology; their combined benefits will be tested in our future devices. We suggested the TMAH surface treatment as an efficient post gate-recess process. It was found that the post-deposition annealing of the Al_2O_3 gate insulator is effective in reducing the charge trapping and hysteresis. To improve the series resistance and uniformity of the threshold voltage, the GaN MOSFET with re-grown n+ GaN source/drain was proposed. The strain-controlled GaN MOSFET features extremely high 2DEG density in source and drain, greatly improving the current drivability in the channel. The MOSFETs with the p-GaN back-barrier exhibited much lower buffer leakage current than devices without the back-barrier. Finally, our preliminary GaN FinFETs with thin and narrow fin had normally-off operation (due to the 2DEG depletion by the lateral gate action) and encouraging mobility.

Table 1. Summary of the techniques explored for the optimization of normally-off GaN MOSFETs: key parameters, advantages, and disadvantages.

Techniques	Key parameters	Advantages	Disadvantages
post-recess TMAH treatment	gate leakage current, effective mobility	reduction of gate leakage current, improved effective mobility	additional process step
post-deposition annealing	V_{th} shift, charge trapping	control of V_{th}, reduction of charge trapping and hysteresis	additional process step
regrowth of S/D	S/D ohmic recess depth, effective mobility	uniformity in V_{th}	growth complexity
strain management	2-DEG density, on-resistance	improvement of the on-current, high effective mobility	growth control
p-GaN back-barrier	buffer leakage current	reduction of buffer leakage current	poor crystalline quality of p-GaN layer
3-D structure	V_{th}, mobility	high V_{th}, high on-current	low fabrication yield

Acknowledgments

This work was supported by 2008 Brain Korea 21 (BK21), the National Research Foundation of Korea (grants 2011-0016222, 2011-0001076), "Survey of high efficiency power devices and inverter system for power grid" project of Korea Ministry of Knowledge Economy, and WCU (World Class University) program (grant R33-10055) and the IT R&D program of MKE/KEIT (10038766, Energy Efficient Power Semiconductor Technology for Next Generation Data Center).

References

1. X. Hu, G. Simin, J. Yang, M. Asif Khan, R. Gaska, and M. S. Shur, *Electronics Lett.*, **36**, 753 (2000).
2. Y. Uemoto, M. Hikita, H. Ueno, H. Matsuo, H. Ishida, M. Yanagihara, T. Ueda, T. Tanaka, and D. Ueda, *IEEE Trans. Electron Devices*, **54**, 3393 (2007).
3. L. Yuan, H. Chen, and K. J. Chen, *IEEE Electron Device Lett.*, **32**, 303 (2011).
4. K. S. Im, J. B. Ha, K. W. Kim, J. S. Lee, D. S. Kim, S. H. Hahm, and J. H. Lee, *IEEE Electron Device Lett.*, **31**, 192 (2010).
5. T. Oka and T. Nozawa, *IEEE Electron Device Lett.*, **29**, 668 (2008).
6. C. Y. Chang, S. J. Pearton, C. F. Lo, F. Ren, I. I. Kravchenko, A. M. Dabiran, A. M. Wowchak, B. Cui, and P. P. Chow, *Appl. Phys. Lett.*, **94**, 263505 (2009).
7. H. Kambayashi, Y. Niiyama, S. Ootomo, T. Nomura, M. Iwami, Y. Satoh, S. Kato, and S. Yoshida, *IEEE Electron Device Lett.*, **28**, 1077 (2007).
8. M. Kanamura, T. Ohki, T. Kikkawa, K. Imanishi, T. Imada, A. Yamada, and N. Hara, *IEEE Electron Device Lett.*, **31**, 189 (2010).
9. B. Lu, O. I. Saadat, and T. Palacios, *IEEE Electron Device Lett.*, **31**, 990 (2010).
10. Y. Q. Wu, T. Shen, P. D. Ye, and G. D. Wilk, *Appl. Phys. Lett.*, **90**, 143504 (2007).
11. M. D. Groner, J. W. Elam, F. H. Fabreguette, and S. M. George, *Thin Solid Films*, **413**, 186 (2002).
12. S. Jakschik, U. Schroeder , T. Hecht , M. Gutshe , H. Seidl, and J. W. Bartha, *Thin Solid Films*, **425**, 216 (2003).
13. C. Ostermaier, H. C. Lee, S. Y. Hyun, S. I. Ahn, K. W, Kim, I. H. Cho, J. B. Ha, and J. H. Lee, *Phys. Stat. Sol. (c)*, **6**, 1992 (2008).
14. G. Ghibaudo, *IEEE Electronics Letters*, **24**(9), 543 (1988).
15. E. Calleja, F. J. Sa´nchez, D. Basak, M. A. Sanchez-Garcia, E. Munoz, I. Izpura, F. Calle, J. M. G. Tijero, J. L. Sanchez-Rojas, B. Beaumont, P. Lorenzini, and P. Gibart, *Physical Review B*, **55**(7), 4689 (1997).
16. U. Kaufmann, M. Kunzer, H. Obloh, M. Maier, Ch. Manz, A. Ramakrishnan, and B. Santic, *Physical Review B*, **59**(8), 5561 (1999).
17. S. M. Hubbard, G. Zhao, D. Pavlidis, W. Sutton, and E. Cho, *J. Cryst. Growth*, **284**, 297 (2005).
18. S. Heikman, S. Keller, S. P. DenBaars, and U. K. Mishra, *Appl. Phys. Lett.*, **81**, 439 (2002).
19. S. Kato, Y. Satoh, H. Sasaki, I. Masayuki, and S. Yoshida, *J. Cryst. Growth*, **298**, 831 (2007).
20. T. Palacios, A. Chakraborty, S. Heikman, S. Keller, S. P. DenBaars, and U. K. Mishra, *IEEE Electron Device Lett.*, **27**, 13 (2006).
21. F. Medjdoub, J. Derluyn, K. Cheng, M. Leys, S. Degroote, D. Marcon, D. Visalli, M. Van Hove, M. Germain, and G. Borghs, *IEEE Electron Device Lett.*, **31**, 111 (2010).
22. S. Karmalkar, J. Deng, M. S. Shur, and R. Gaska, *IEEE Electron Device Lett.*, **22**, 373 (2001).
23. H. C. Lee, S. Y. Hyun, H. I. Cho, C. Ostermaier, K. W. Kim, S. I. Ahn, K. I. Na, J. B. Ha, D. H. Kwon, C. K. Hahn, S. H. Hahm, H. C. Choi, and J. H. Lee, *Jpn. J. Appl. Phys.*, **47**, 2824 (2008).
24. D. S. Kim, K. S. Im, H. S. Kang, K. W. Kim, S. B. Bae, J. K. Mun, E. S. Nam, and J. H. Lee, *Jpn. J. Appl. Phys.*, **51**, 034101 (2012).
25. S. Liu, Y. Cai, G. Gu, J. Wang, C. Zeng, W. Shi, Z. Feng, H. Qin, Z. Cheng, C. Chen, and B. Zhang, *IEEE Electron Device Lett.*, to be published (2012).
26. B. Lu, E. Matioli, and T. Palacios, *IEEE Electron Device Lett.*, to be published (2012).
27. T. Zimmermann, Y. Cao, X. Luo, D. Jena, and H. Xing, in *Proc. 67th Annu. Device Res. Conf.*, 129 (2009).
28. K.-W. Kim, S.-D. Jung, D.-S. Kim, H.-S. Kang, K.-S. Im, J.-J Oh, J.-B. Ha, J.-K. Shin, and J.-H. Lee, *IEEE Electron Device Lett.*, **32**, 1376 (2011).

SILICON-ON-INSULATOR MESFETs AT THE 45nm NODE

WILLIAM LEPKOWSKI and SETH J. WILK

SJT Micropower Inc., 16411 N. Skyridge Lane
Fountain Hills, AZ 85268-1515, USA
w.lepkowski@sjtmicropower.com

M. REZA GHAJAR, ANURADHA PARSI and TREVOR J. THORNTON

School of Electrical, Computer and Energy Engineering, Arizona State University
PO Box 876206, Tempe, AZ 85287-6206, USA
t.thornton@asu.edu

Metal-semiconductor field-effect-transistors (MESFETs) have been fabricated using a commercially available 45nm silicon-on-insulator (SOI) CMOS foundry with no changes to the process flow. Depending upon the layout dimensions, these n-channel, depletion mode devices can be designed for high current drive ($I_D^{SAT} \geq 100mA/mm$), high operating frequency ($f_{max} >35$ GHz) or enhanced breakdown voltage ($V_{BD} >25V$). The design flexibility provided by the SOI MESFETs, coupled with the high performance of ULSI CMOS at the 45nm node will enable a variety of analog, RF and mixed signal applications.

Keywords: Silicon on Insulator; MESFETs; Spice models.

1. Introduction

The first CMOS-compatible SOI MESFETs fabricated with no changes to the process flow were made using a 0.8μm CMOS process and were presented at the Workshop on Frontiers in Electronics (WOFE) in 2004 [1]. Since then the MESFETs have been demonstrated at four different foundries, using technology nodes from 350nm to 150nm. The MESFETs can be seamlessly integrated with either partially-depleted [2-4] or fully-depleted [5, 6] SOI (silicon-on-insulator) technologies with characteristics that augment those of the ULSI CMOS. Very recently a 45nm PD-SOI CMOS foundry has been used to fabricate SOI MESFETs with the highest f_{max} and current drive reported to date. Like those reported earlier, the latest n-channel MESFETs demonstrate depletion mode operation with threshold voltages of approximately -0.6V. Unlike the accompanying CMOS which depends upon the integrity of a fragile gate oxide, the gate of the SOI MESFETs is based on a Schottky contact that can tolerate substantial current flow under forward and reverse bias. As a result, the breakdown voltage of the MESFETs is substantially higher than that of the CMOS devices. In fact, the breakdown voltage, V_{BD}, of the MESFET can be tailored by choosing the length of the source and drain extensions at either end of the gate, and $V_{BD} >25V$ has been demonstrated, albeit with reduced current drive, and lower f_T and f_{max}.

The high breakdown voltage of the SOI MESFETs and their good RF performance offers a number of interesting CMOS-MESFET integrated circuit design opportunities.

DC power management applications such as low dropout linear regulators are a good example. In [7] an SOI MESFET was configured as a source follower pass transistor controlled by a CMOS error amplifier to provide stable voltage regulation under all load conditions. Integrated CMOS-MESFET analog building blocks have been described in [6]. For RF applications a cascode architecture with a MOSFET as the common-source device and a MESFET as the common-gate is attractive for high efficiency power amplifiers that can tolerate wider voltage excursions than the baseline CMOS devices alone.

This paper presents preliminary results from three foundry runs that have been completed using the same 45 nm SOI CMOS foundry. The DC and RF characteristics of the MESFETs are described with particular emphasis on the design trade-offs between current-drive, switching speed and breakdown voltage. Process monitor blocks have been included in each tapeout to better understand the manufacturability of the MESFETs at the 45nm node. Statistical data showing the run-to-run variations in threshold voltage, current drive and transconductance has been collected. The results confirm very acceptable manufacturing tolerances. The paper is presented as follows: Section 2 describes prior work on silicon based MESFETs and summarizes recent developments using SOI technologies. This is followed in Section 3 by measurements of the DC behavior, and breakdown characteristics of the MESFETs. Section 4 describes the RF characteristics and the trade-offs between operating frequency and breakdown voltage. The statistical variations observed during two different process runs are described in Section 5. Section 6 concludes with some potential applications of the MESFET devices and directions for future work.

2. Silicon Based MESFETs

MESFETs have a long history with the earliest suggestion of the device being a patent issued to Lilienfield in 1926. It was a number of years later, however, before the first practical demonstration of a MESFET was presented by Mead in 1966 using GaAs as the semiconductor channel [8]. The first demonstration of a silicon based MESFET was published by Drangeid in 1968 [9] using an epitaxial arsenic doped channel layer grown on a high resistivity p-type substrate. The device used a Au-Cr-Ni metal gate 1 μm in length and had a maximum operating frequency of 6 GHz, which at the time was considerably faster than values reported for MOSFETs. However, the rapid progress with CMOS and the superior transport properties of GaAs meant that commercial MESFET development focused primarily on compound semiconductor materials. As a result, the available literature on silicon based MESFETs is relatively small. With the development of ion implantation it was possible to form n-channels on lightly doped p-type substrates without the need for silicon epitaxy. Early work focusing on Si MESFETs for LSI logic used PtSi gate electrodes of length 2 μm patterned by contact printing [10] and as small as 0.25 μm using electron beam lithography [11]. Optimization of the channel doping to minimize substrate parasitics increased f_{max} to 14GHz [12]. A complementary Si

MESFET technology using bulk CMOS substrates was demonstrated by MacWilliams and Plummer [13].

With the introduction of silicon-on-insulator (SOI) substrates (including silicon-on-sapphire, SOS) interest in SOI MESFETs began to grow. One of the first demonstrations used an SOS substrate to produce both enhancement and depletion mode devices with drawn gate lengths in the range $0.2 - 1$ µm [14]. Complementary SOS MESFETs were fabricated by Magnusson et al. [15] with a focus on sub-threshold operation. The early SOS substrates suffered from reduced electron mobility due to defects at the silicon: sapphire interface. With the introduction of SOI wafers, MESFET channel layers with higher electron mobility became available, leading to enhanced transconductance compared to earlier devices [16, 17]. The use of SOI substrates eliminated many of the problems associated with off-state leakage currents allowing for near-ideal sub-threshold performance [18].

Along with the experimental development, a growing body of work addresses the simulation and modeling of SOI MESFETs. Early work addressed the limits of silicon MESFET scaling [19]. Both analytical [20-22], and numerical models [23-25] have been developed to aid in the design of SOI MESFETs. Recently, Spice models for circuit simulation have been developed [26, 27] based on the widely used TOM3 model that was originally developed for GaAs MESFETs.

Current SOI CMOS technologies use silicides for the low resistance contacts to the source, gate and drain contacts. The silicides make almost ideal Schottky contacts when deposited on lightly doped n-well regions and by appropriate use of a silicide block mask the Schottky gate can be separated from the low resistance silicide contacts on the source and drain as described in [1]. This patterning of silicide above n-wells often violates the 'electrical rules' of the CMOS design kit but does not change the process flow itself or affect the operation of the CMOS. A more challenging issue in using highly scaled CMOS for MESFET fabrication is the introduction of SiGe layers in the source and drain regions to enhance the channel mobility through strain engineering. The SiGe regions are formed using masking layers that are derived from the designed layout without being drawn directly. As a result the silicide on n-well active regions will be formed above epitaxial SiGe by default. To avoid this outcome a mechanism that allows some control over the derived layers needs to be available from the foundry. The minimum feature sizes of the MESFET are ultimately controlled by the layout rules of the CMOS design kit. For example the minimum gate length of the MESFET is controlled by the design rule that specifies the closest allowable separation between silicide block regions, labeled as 'spacer' in Fig. 1.

The same basic approach for SOI MESFET fabrication described in [1] and summarized above has been used in a range of CMOS process nodes from 0.8 µm down to the latest demonstration described here for the 45 nm node. Interestingly, despite continuing reduction in silicon channel thickness, increased well doping and changes in silicide thickness from one node to the next, the MESFET threshold voltage has stayed remarkably constant in the range -0.5V to -0.6V. Of course, reductions in the minimum feature size with each node has allowed the MESFET gate length to be scaled with corresponding increases in current drive, transconductance and switching frequency. However, since the CMOS process flow is designed to optimize MOSFET performance the MESFETs suffer from short channel effects at gate lengths that are typically less than 2-3X longer than the shortest available CMOS devices. For this MESFET demonstration at the 45 nm node, conservative design rules were used, resulting in MESFETs with minimum gate lengths of 200 nm.

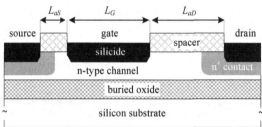

Fig. 1. Schematic cross-section of the SOI MESFET. The minimum length of the gate, L_G, and the access regions L_{aS} and L_{aD} are controlled by the design rules that stipulate the minimum size and separation between the silicide block spacer layers formed by a deposited dielectric.

3. D.C. Characteristics of SOI MESFETs

A series of MESFET devices with minimum gate lengths of 200nm were designed with source and drain access regions ranging in length from 200nm to 2μm. The goal of this study was to understand the trade-offs between DC current drive and operating frequency on the one hand and the off-state breakdown voltage on the other. Previous work [3] has shown that the principal breakdown mechanism in these devices is due to avalanche enhanced reverse gate current at the drain end of the channel. By increasing the access lengths, especially L_{aD}, the electric field between the edge of the gate and the drain contact can be reduced for a given drain voltage, thereby allowing higher voltage operation. This is illustrated in Fig. 2 which shows the family-of-curves for a MESFET with $L_G=L_{aS}=L_{aD}=200$ nm (Fig. 2a) and for another with the same gate length but with $L_{aS}=1μm$ and $L_{aD}-1μm$ (Fig. 2b).

The turn-on characteristics of both devices are shown in Fig. 3 with a peak current drive approaching 90 mA/mm for the device with $L_{aD}=200$nm. As expected the device with longer L_{aS}, L_{aD} has lower current drive and a reduced gate current due to the increased parasitic resistances in the device with longer access regions.

It is clear from Fig. 2b that by appropriate choice of the source and drain access lengths, the off-state breakdown voltage of MESFETs can be designed to exceed 20V. To quantify the variation of breakdown voltage with the dimensions L_{aS} and L_{aD}, the drain current injection technique was used as described in [3]. In this approach a fixed drain current bias corresponding to 1 mA/mm is applied to the device while reducing the gate voltage from 0V to close to threshold. As the device is switched off the drain voltage has

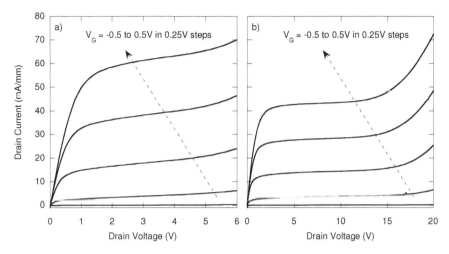

Fig. 2. Family-of-curves measurements of 200nm gate length MESFETs fabricated using the 45nm process. a) shows measured results for devices with $L_{aD} = L_{aS}$ of 200nm while b) shows the higher voltage device with $L_{aD} = L_{aS}$ of 1µm.

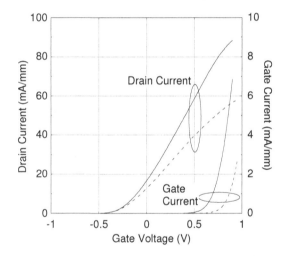

Fig. 3. The turn-on characteristics for both of the MESFETs in Fig. 2. The solid lines are for the device with $L_{aS}=L_{aD}=200$nm while the dashed lines are for the device with $L_{aS}=L_{aD}=1$µm. The drain voltage was fixed at 2V.

to increase to maintain the fixed drain current. At some point the drain voltage reaches the value at which avalanche breakdown occurs from gate to drain. The drain current is then supported by flowing out of the gate contact which causes the drain voltage to increase less rapidly. These trends are displayed in Fig. 4 which shows drain current injection measurements for a device with $L_{aS}=500$nm and $L_{aD}=2$µm. It can be seen that as the gate voltage approaches threshold a significant fraction of the drain current is flowing out of the gate contact. This condition is reversible with no damage to the device provided the gate and drain currents are carefully limited. The soft breakdown voltage is defined as the intersection between the two asymptotes present in the drain voltage curve as indicated by the two dashed lines in Fig. 4. It is clear from the data that the device can survive higher drain voltages but using the intersection between the two asymptotes to define the breakdown voltage is a convenient method that can be applied to compare results from different devices.

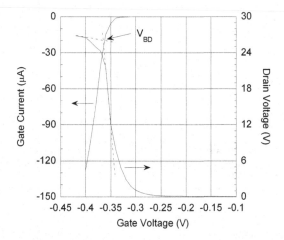

Fig. 4. Data used to extract the soft breakdown voltage of an SOI MESFET with L_G = 200 nm, L_{aS} =500nm and L_{aD} =2μm. For this device the gate width is 300 μm and the device is biased with a fixed drain current of 300 μA corresponding to 1 mA/mm. As described in the text the off-state breakdown voltage is determined from the intersection of the two dashed lines and for this device V_{BD} ~ 26V.

The drain current injection technique was applied to MESFETs with the same gate lengths of 200 nm but with varying L_{aS} and L_{aD} dimensions. The extracted breakdown voltages are shown in Fig. 5. As expected the breakdown voltage increases with increasing L_{aD}.

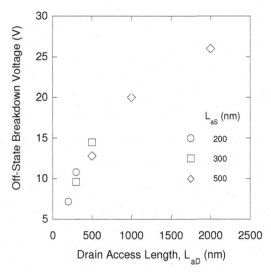

Fig. 5. The off-state breakdown voltage of 200nm gate length MESFETs with varying L_{aS} and L_{aD}.

4. R.F. Characteristics of SOI MESFETs

MESFETs with 200nm gate lengths were configured with ground-signal-ground (GSG) pads for on-wafer probing of S-parameters using an Agilent 8510C network analyzer. The effects of cable loss and pad parasitics were accounted for by careful de-embedding. Figure 6 shows the measured forward current gain, h_{21}, and maximum available gain (MAG) as a function of frequency for a device with $L_G = L_{aS} = L_{aD} =200$nm. By extrapolating h_{21} and MAG to 0 dB we extract $f_T =25$ GHz and $f_{max} = 35$GHz, respectively.

For devices with a fixed gate length (here $L_G = 200$ nm) the values of f_T and f_{max} depend upon bias conditions and the length of the access regions. Figure 7 shows the trends in f_T and f_{max} for two devices that span a wide range of L_{aD}. For the device with $L_{aS} = L_{aD} = 200$nm the maximum operating frequency, f_{max}, is greater than the cut-off frequency, f_T. However, for the device with $L_{aS} =L_{aD} = 1000$ nm, the additional parasitic resistance introduced by the increased access lengths results in $f_{max} \sim f_T$.

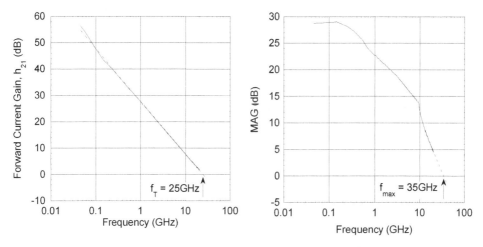

Fig. 6. a) The forward current gain, h_{21}, and b) maximum available gain, MAG, as a function of frequency after pad de- embedding. The dashed lines are 20dB/decade extrapolations of the measured data to extract the values of f_T and f_{max}. The bias conditions are $V_G=0.25$V and $V_D =2$V for a device with $L_G= L_{aS} = L_{aD} =200$ nm.

The trade-off between peak cut-off frequency and breakdown voltage is shown in Fig. 8. The data has been extracted from four different devices each with gate length of 200 nm but different values of L_{aS} and L_{aD}.

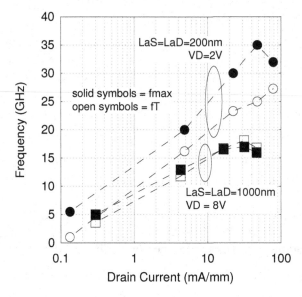

Fig. 7. The de-embedded values of f_T and f_{max} as a function of drain current for two different MESFETs.

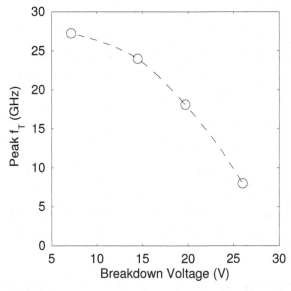

Fig. 8. The peak cutoff frequency of four devices with a range of breakdown voltages due to differences in drain and source access lengths.

5. MESFET Manufacturability

A MESFET process monitor (PM) block has been developed for the 45 nm technology and included as part of two separate designs that were submitted to the foundry. The second design (Run 2) was submitted 4 months after the first design (Run 1). The PM block MESFETs used a somewhat more aggressive application of the design rules to achieve devices with gate lengths of 180 nm and access regions with equal lengths of 100 nm. The channel width of the PM MESFETs was 100 μm. Test structures on each PM block have been measured with the purposes of understanding the run-to-run variation in key device parameters. A DC probe station was used to sweep the MESFET bias to collect data from which the threshold voltage and current drive was extracted. Figure 9a shows the measured Gummel plot. For simplicity the threshold voltage has been defined as the value of the gate voltage for which $I_D = 10^{-4}$ A as indicated by the arrow in Fig. 9a. In a similar fashion the family of curves (Fig. 9b) is used to extract the current drive, defined as the drain current measured for a bias of $V_D = 4$V and $V_G = +0.5$V and indicated by the open circle in Fig. 9b.

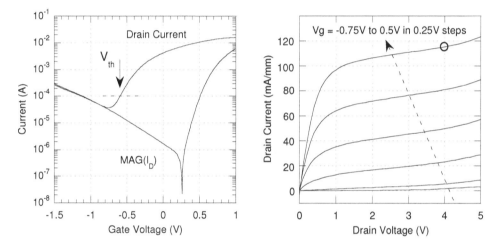

Fig. 9. a) The Gummel plot measured for a typical PM MESFET device at a fixed drain voltage of 2V. The threshold voltage is defined as the value of gate voltage for which $I_D=10^{-4}$ A as indicated by the dashed line. b) The PM MESFET family of curves with the current drive extracted from the value of drain current when $V_D=4$V and $V_G=+0.5$ V as indicated by the open circle.

Using the methods described above the threshold voltage and current drive have been extracted. The threshold voltage distribution for 31 devices is shown in Fig. 10. The data from Run 1 is shown in Fig. 10a and can be compared to the same measurements from Run 2 shown in Fig. 10b. The histogram for the entire data set (Runs 1 and 2) is presented in Fig. 10c. The total number of devices used for threshold extraction is still relatively small but the preliminary results are very encouraging with good threshold control over the two process runs. From the entire data set the mean threshold voltage is -0.63V with a standard deviation of 35 mV, corresponding to a 5.6% variation.

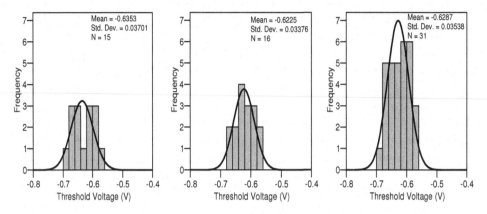

Fig. 10. The threshold voltage distributions for (a) Run 1 and (b) Run 2 submitted four months later. The distribution across all 31 devices is shown in (c).

The variation in current drive is shown in a similar fashion in Fig. 11. Here, data from 60 devices has been extracted and the mean current drive across both process runs is 102 mA/mm with a standard deviation of 13 mA/mm corresponding to a variation of 13%. Based on the data in Figs. 10 and 11 it appears that the run-to-run variation between the MESFET PM structures is similar to the die-to-die variation within the same processed batch of wafers. This conclusion is encouraging because it suggests that MESFET fabrication through the 45nm foundry is well behaved and will support a manufacturing supply chain for high-volume, low-cost MESFET based integrated circuits.

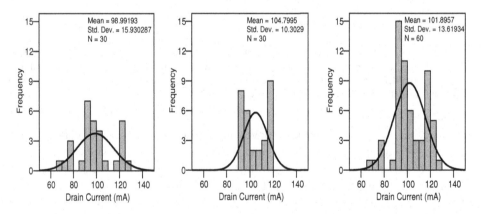

Fig. 11. The current drive distributions for (a) Run 1 and (b) Run 2. The distribution across all 60 devices is shown in (c).

6. Future Directions

The SOI MESFETs described above offer a number of capabilities that augment those of the CMOS technology node used for the fabrication. Chief amongst these is the enhanced voltage operation of the MESFETs. With continued CMOS scaling, the maximum supply voltage of the standard CMOS logic devices is approaching 1V or less and the MESFETs offer a number of choices for higher voltage input/output devices. Analog circuits with inductive elements can expose active devices to voltages that are 2-4X higher than those of the supply rails. These include DC-DC power converters and certain classes of switching amplifiers. By using the MESFET as the pass transistor of a boost or buck converter it can be used to control the high voltage switching waveforms while the integrated CMOS performs the required logic and control functions at lower voltages. A similar approach can be used for high efficiency RF power amplifiers (e.g. Class E etc.) in which the MESFET controls the switching waveforms with amplitudes that exceed $2V_{DD}$. The high voltage capability of the MESFET is also beneficial for R.F. amplifiers due to the reduced constraints it places on the output matching circuits. For highly scaled CMOS with maximum operating voltages of ~1V, it is necessary to use wide devices to achieve the currents required for even modest power amplifiers with P_{out}~1W. This results in low transistor output impedance of e.g. 5Ω, that require a 10:1 impedance transformation in the output matching network. Such high transformation ratios lead to lower power added efficiency. In contrast, a MESFET biased at V_{DD}=10V can use 10X lower currents to achieve the same RF power, bringing the output impedance closer to 50Ω, leading to increased PAE. Examples of two integrated CMOS-MESFET circuit opportunities are described below followed by a layout scheme that will allow MESFET based circuits to be fabricated using FD-SOI CMOS technologies.

6.1. *MESFET pass transistor for low dropout linear regulators*

The MESFET architecture offers additional design opportunities beyond the enhanced voltage capability. The fact that it is a depletion mode, n-channel device means it can be used in linear regulator applications without the need for an external compensation capacitor. Such a design is shown in Fig. 12 and has been demonstrated by Lepkowksi et al. [28, 7]. In this approach the MESFET pass transistor is configured as a voltage-follower forming a single-pole system that is intrinsically stable under all load conditions. In contrast, a CMOS only approach typically use a p-channel, enhancement mode device as the output pass transistor of a two-pole system requiring an output capacitance with a narrow range of equivalent series resistance to ensure stability. The circuit of Fig. 12 makes use of a CMOS folded cascode op-amp with an output buffer stage that is required to drive the gate current of a large MESFET pass transistor. As described in [29] the MESFET pass transistor is capable of supporting load currents up to 1A with a dropout voltage of <170mV.

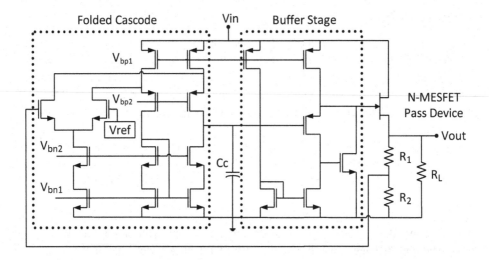

Fig. 12. Circuit schematic of a MESFET based linear regulator with integrated CMOS error amplifier.

6.2. *MOSFET- MESFET cascode for enhanced voltage RF power amplifiers*

The low breakdown voltage of ULSI CMOS is especially limiting for RF power amplifier (PA) applications. One approach to increase the permissible voltage swing at the output of the PA is to use a cascode architecture as shown in Fig. 13. A high frequency MOSFET is used as the common-source (CS) input device while a high voltage MESFET is used as the common-gate (CG) output. By suitable choice of transistor widths and bias points the majority of the voltage swing is dropped across the CG device. A similar approach has been used to enhance the voltage range of highly scaled SiGe HBT RF amplifiers [30].

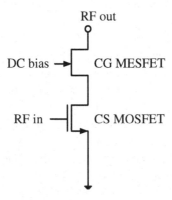

Fig. 13. Circuit schematic of a MOSFET-MESFET cascode RF power amplifier that makes use of a high voltage MESFET as the common-gate device to allow larger RF output voltage swings.

6.3. *Scaling SOI MESFETs to fully-depleted SOI CMOS*

It is important to consider how the MESFET architecture will scale as SOI CMOS migrates from partially-depleted (PD) to fully-depleted (FD) technologies. The architecture of Fig. 1 has worked well in a variety of PD-SOI foundries from 0.8μm to the latest demonstration at the 45nm node, and is expected to migrate well to the 32 nm node. However, with FD-SOI CMOS the silicon channel thickness will be too small to support depletion mode operation under the silicide gate. At this point a laterally depleting MESFET will be required as demonstrated in [5]. A 3D schematic view of the FD-SOI MESFET is shown in Fig. 14a. Islands of silicide separate silicon channels that remain under the silicide block regions. The depletion layer formed by the silicide Schottky barrier can be used to control the currents flowing through the silicon channels by lateral confinement as indicated in Fig. 14b. In this fashion both n- and p-channel devices have been demonstrated [5] with threshold voltages that can be tuned by adjustments to the physical channel width, L_{CW}. As a result, silicon MESFET based circuits can be envisaged to the end of the SOI CMOS roadmap offering mixed-signal and analog circuit design options that are not readily available from a CMOS-only approach.

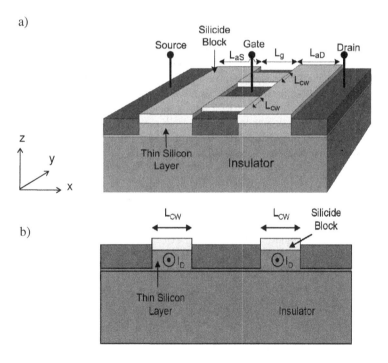

Fig. 14. a) 3D representation of the FD-SOI MESFET described in [5]. Lateral depletion is used to confine the current flow through narrow silicon channels of physical width L_{CW}. b) Cross-section through the silicon channels in the y-z plane.

7. Conclusions

A review of SOI MESFETs fabricated using a 45nm CMOS foundry has been presented. The results confirm the anticipated improvements in DC and RF performance for MESFETs from the most highly scaled CMOS technology used to date. The trade-offs between current drive, RF performance and breakdown voltage have been explored. As with earlier demonstrations the MESFET breakdown voltage at the 45nm node greatly exceeds that of the baseline CMOS devices. Two examples of integrated CMOS-MESFET circuits have been presented to illustrate how the properties of the MESFET can be used to augment the capabilities of the ULSI CMOS. With the ability to fabricate the MESFET architecture using FD-SOI technologies it seems possible that it can be scaled even further than allowed by the 45nm demonstration presented here.

Acknowledgments

Access to the IBM 45nm SOI CMOS technology was provided through the Trusted Access Programs Office (TAPO) and with support from a DARPA STTR contract # W31P4Q-10-C-0020. The development of the MESFET technology for linear regulator applications was supported by NASA through SBIR contract # NNX11CC01C.

References

1. J. Yang, A. Balijepalli, T. J. Thornton, J. Vandersand, B. J. Blalock, M. Mojarradi, and M. E. Wood, "Silicon-Based Integrated MOSFETs and MESFETs: A New Paradigm for Low Power, Mixed Signal, Monolithic Systems using Commercially Available SOI," *International Journal of High Speed Electronics and Systems*, **16**, 723–732 (2006).
2. J. Spann, V. Kushner, T. J. Thornton, J. Yang, A. Balijepalli, H. J. Barnaby, X. J. Chen, D. Alexander, W. T. Kemp, S. J. Sampson, and M. E. Wood, "Total Dose Radiation Response of CMOS Compatible SOI MESFETs," *IEEE Trans. Nuc. Sci.*, **52**, 2398–2402 (2005).
3. J. Ervin, A. Balijepalli, P. Joshi, V. Kushner, J. Yang, and T. J. Thornton, "CMOS-Compatible SOI MESFETs with High Breakdown Voltage," *IEEE Trans. Electron Devices*, **53**, 3129–3135 (2006).
4. A. Balijepalli, R. Vijayaraghavan, J. Ervin, J. Yang, S. K. Islam, and T. J. Thornton, "Large-signal modeling of SOI MESFETs," *Solid-State Electronics*, **50**, 943–950 (2006).
5. W. Lepkowski, J. Ervin, S. Wilk, and T. J. Thornton, "SOI MESFETs Fabricated Using Fully Depleted CMOS Technologies," *IEEE Electron Device Letts.*, **30**, 678–670 (2009).
6. S. Kim, W. Lepkowski, T. J. Thornton, and B. Bakkaloglu, "CMOS compatible high-voltage compliant MESFET-based analogue IC building blocks," *Electronics Letters*, **45**, 624–625 (2009).
7. W. Lepkowski, S. Wilk, B. Bakkaloglu, P. S. Fechner, and T. J. Thornton, "Ultra-low dropout linear regulator using an SOI MESFET," presented at *IEEE International SOI Conference*, San Diego, CA (2010).
8. C. A. Mead, "Schottky barrier gate field effect transistor," *Proceedings of the IEEE*, **54**, 307–308 (1966).
9. K. E. Drangeid, R. Jaggi, S. Middlehoek, T. Mohr, A. Moser, G. Sasso, R. Sommerhalder, and P. Wolf, "Microwave Silicon Schottky-Barrier Field-Effect Transistor," *Electronics Letters*, **4**, 362–363 (1968).

10. H. Muta, S. Suzuki, K. Yamada, Y. Nagahashi, T. Tanaka, H. Okabayashi, and N. Kawamura, "Femto Joule logic circuit with enhancement-type Schottky barrier gate FET," *IEEE Trans. Electron Devices*, **23**, 1023–1027 (1976).

11. H. M. Darley, T. W. Houston, and G. W. Taylor, "Fabrication and performance of submicron silicon MESFET," *IEEE Int. Electron Devices Meeting*, **24**, 62–65 (1978).

12. G. Fernholz and H. Beneking, "Ion implanted Si MESFET's with high cutoff frequency," *IEEE Trans. Electron Devices*, **30**, 837–840 (1983).

13. K. P. Macwilliams and J. D. Plummer, "Device Physics and Technology of Complementary Silicon MESFETs for VlSI Applications," *IEEE Trans. Electron Devices*, **38**, 2619–2631 (1991).

14. J. Nulman and J. P. Krusius, "2-GHz 150 microWatt self-aligned Si MESFET logic," *IEEE Electron Device Letts.*, **5**, 159–161 (1984).

15. U. Magnusson, J. Tiren, H. Norde, and H. Bleichner, "Subthreshold Behavior of Silicon MESFETs on SOS and Bulk Silicon Substrates," *Solid-State Electronics*, **32**, 931–934 (1989).

16. D. P. Vu and A. Sono, "Self-Aligned Si MESFETs Fabricated in Thin Silicon-on-Insulator Films," *Electronics Letters*, **23**, 354–355 (1987).

17. H. Vogt, G. Burbach, J. Belz, and G. Zimmer, "MESFETs in Thin Silicon on SIMOX," *Electronics Letters*, **25**, 1580–1581 (1989).

18. J. Yang, J. Y. Spann, R. Anderson, and T. J. Thornton, "High-frequency performance of sub-threshold SOI MESFETs," *IEEE Electron Device Letts.*, **25**, 652–654 (2004).

19. J. D. Marshall and J. D. Meindl, "An Analytical Two-Dimensional Model for Silicon MESFETs," *IEEE Trans. Electron Devices*, **35**, 373–383 (1988).

20. P. Chattopadhyay, "The DC Characteristics of a Silicon-on-Insulator Metal-Semiconductor Field Effect Transistor," *Semiconductor Science and Technology*, **13**, 1036–104 (1998).

21. T. K. Chiang, Y. H. Wang, and M. P. Houng, "Modeling of threshold voltage and subthreshold swing of short-channel SOI MESFET's," *Solid-State Electronics*, **43**, 123–129 (1999).

22. P. Hashemi, A. Behman, E. Fathi, A. Afzali-Kusha, and M. El Nokali, "2-D Modeling of Potential Distribution and Threshold Voltage of Short Channel Fully Depleted Dual Material Gate SOI MESFET," *Solid-State Electronics*, **49**, 1341–1346 (2005).

23. C. S. Hou and C. Y. Wu, "A design strategy for short gate length SOI MESFETs," *Solid-State Electronics*, **39**, 361–367 (1996).

24. T. J. Thornton, "Schottky junction transistor-micropower circuits at GHz frequencies," *IEEE Electron Device Letts.*, **22**, 38–40 (2001).

25. T. J. Thornton, "Physics and applications of the Schottky junction transistor," *IEEE Trans. Electron Devices*, **48**, 2421–2427 (2001).

26. A. Balijepalli, R. Vijayaraghavan, J. Ervin, J. Yang, S. K. Islam, and T. J. Thornton, "Large-signal modeling of SOI MESFETs," *Solid-State Electronics*, **50**, 943–950 (2006).

27. A. Balijepalli, J. Ervin, W. Lepkowski, Y. Cao, and T. J. Thornton, "Compact modeling of a PD SOI MESFET for wide temperature designs," *Microelectronics Journal*, **40**, 1264–1273 (2009).

28. W. Lepkowski, S. Wilk, S. Kim, B. Bakkaloglu, and T. J. Thornton, "A Capacitor-Free LDO Using a FD Si-MESFET Pass Transistor," presented at *52nd IEEE International Midwest Symposium on Circuits and Systems* (MWSCAS), Cancun, Mexico, 2-5 August (2009).

29. W. Lepkowski, S. J. Wilk, M. Reza Ghajar, B. Bakkaloglu and T. J. Thornton "An Integrated MESFET Voltage Follower LDO for High Power and PSR RF and Analog Applications" to be presented at the *IEEE Custom Integrated Circuits Conference*, San Jose, CA, Sept. 9-12 (2012).

30. J. Andrews, J. D. Cressler, W-M L. Kuo, C. Grens, T, Thrivikraman and S. Phillips "An 850 mW X-Band SiGe Power Amplifier" pp. 109-112, Proc. of the *IEEE BiCMOS Circuits and Technology Meeting* (BCTM), Monterey, CA, 13-15 Oct. (2008).

ADVANCED CONCEPTS FOR FLOATING-BODY MEMORIES

FRANCISCO GÁMIZ*, NOEL RODRIGUEZ* and SORIN CRISTOLOVEANU†

*Dept. Electrónica y Tecnología de Computadores, University of Granada, 18071 Granada, Spain
fgamiz@ugr.es, noel@ugr.es
†IMEP-INP-Grenoble Minatec, 3 Parvis Louis Néel, BP 257, 38016 Grenoble Cedex 1, France
sorin@enserg.fr

With 30nm-class memory cells in production and 20nm-class (20-29nm feature-size) memory targeted for next year, the standard 1-Transistor + 1-Capacitor (1T+1C) DRAM industry is making prominent efforts to improve the scalability of the cell capacitor while maintaining the minimum capacitance requirements for state discrimination, immune to noise (C~25fF/cell). To achieve the capacitance requirement, the DRAM cell has evolved from its initial planar implementation to complex three-dimensional structures. The increment in complexity and the large difference in size between the transistor and capacitor of each cell have motivated the search for Floating-Body Single-Transistor DRAM (1T-DRAM). The underlying idea behind 1T-DRAMs is the development of single-device memory cells with a pronounced hysteresis effect and fast operation. This chapter is focused on the floating-body effect as a primary source of hysteresis. We present new concepts able to deal with the basic limitations of 1T-DRAM while maintaining its simplicity. The floating-body 1T-DRAMs can be reconciled with the aggressive scaling constrains by considering new ideas which make possible the coexistence of electron and holes in the same ultrathin transistor. The best approach is to isolate each type of carrier in an specific potential well which is not created specifically by the bias conditions (unlike standard 1T-DRAMs) but by the physical structure of the device.

Keywords: Dynamic Random Access Memory (DRAM); Single-Transistor DRAM (1T-DRAM); floating body effect; silicon on insulator (SOI); band-to-band tunneling; embedded DRAM; low-power memory; fast memory devices.

1. Introduction

Since the birth of modern semiconductor industry, the need to store temporarily digital information (to feed the microprocessors with data and instructions) has grown in parallel with the need for higher computational performance. During the last three decades, two major sets of storage cells have fulfilled the requirement for temporal memory banks in the information technology world: the Static Random Access Memory (SRAM) and the Dynamic Random Access Memory (DRAM) [1].

SRAM cells are formed by six transistors: four of them constitute a latch of two inverters, and the other two transistors allow the access (reading and writing) from and into the cell. Major advantages of SRAMs are: full compatibility with the CMOS process, robustness to variability issues present in the fabrication technology as well as very high operation speed. However the penalty to pay is large area consumption per cell, which nowadays is an important drawback as compared with other technologies. This complexity limits the amount of data that a single chip can store, i.e., SRAM chips cannot hold as many bytes as DRAM chips. Typically SRAM cells are found in buffer and cache memory of microprocessors [1].

The second major set of volatile semiconductor memories is the DRAM. The basic principle of operation has remained unaltered for more than 30 years [2], but the structure of the cell has evolved from a planar layout to complicated three-dimensional structures, including high-k materials and corrugated capacitor surfaces [3]. The increasing demand for DRAMs has resulted in one of the most vivid markets of the semiconductor industry. However, serious constraints are at present threatening the DRAM supremacy, as industry continues pushing the dimensions of the semiconductor devices towards the decananometer range [4]:

(1) The cell capacitor should be able to store enough charge to allow the discrimination, immune to noise, between the two memory states. Increasing or maintaining the capacitance size is in contradiction to scalability because of the high area consumption. The gap in terms of real estate between the active device (the transistor which is continuously shrunk) and the capacitor increases, the latter being the limiting factor.
(2) The charge must be transferred with a minimum delay from and into the capacitor.
(3) Both the capacitor and the transistor junction should have a very low leakage current to avoid compromising the retention time.

The external storage element is a burden for the DRAM survival. The step forward that each technology node requires in terms of technological, material and design advances, represents a paramount challenge for the integration of the capacitor inside the cells. This is the main reason of the increasing momentum that the research of new memory cells, free of capacitors, has gained during the last decade.

1T-DRAM is a wide concept that includes a set of memories intended to be potential substitutes of the standard DRAM technology. All of them have a common feature: they avoid using any external storage element. The memory cell consists of a single device (transistor or transistor-like) where the information is stored, i.e., the same device is used to store the information and to read it. Within the 1T-DRAM family, a vast collection of structures has been accommodated: from single transistors to more complicated thyristor-like structures [5].

In this chapter we focus on a particular set belonging to the 1T-DRAM family: the so-called Floating-Body (FB) 1T-DRAM cells.

2. First Generation: Partially Depleted FB 1T-DRAM

About 20 years ago, the pronounced hysteresis effect in Partially-Depleted Silicon-On-Insulator (PD-SOI) transistors has been an intense matter of investigation [6-9]. Since any kind of hysteresis in a transistor entails a memory effect, much attention started to be paid to the design and fabrication of new memory cells on SOI substrates [10-11].

   The particularity of the PD-SOI technology is originated by the floating-body effect [12,13]. The transistor body in PD-SOI devices is isolated from any contact, remaining floating from an electrical point of view and neutral from a carrier balance point of view.

   The research of the floating-body effect in PD-SOI devices rose exponentially when at the beginning of the last decade the start-up company, Innovative Silicon, introduced the Z-RAM memory cell [14]. The basic principle of operation of the cell was the shift induced in the threshold voltage of the PD-SOI MOSFET, caused by the injection of holes in the Floating-Body (a transient overpopulation of holes in the Floating-Body) [15]. This transitory shift of the threshold voltage leads to two different current levels at a given bias point (Fig. 1) [14]. At equilibrium (stable state), the Floating-Body of the transistor remains neutral (Fig. 1a); this situation defines the '0' state. The '1' state is forced by charging the body with holes produced by impact ionization mechanism [13] that occurs when a large current flows through the device. The consequence of the stored charge is an increase in the potential of the body of the device and a decrease in the threshold voltage: for the same bias a larger current is then obtained.

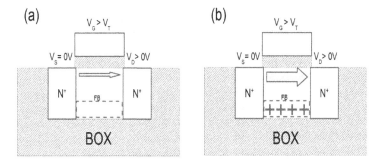

Fig. 1. Schematic operation of a Z-RAM memory cell. (a) In the stable state '0', the floating body remains neutral (dotted region). (b) The '1' state is forced by charging the floating body with an excess of holes, which makes the threshold voltage of the transistor to decrease.

   Figure 2 presents simulation results obtained, under the drift-diffusion approximation, with calibrated models for impact ionization. The topside of the figure shows an example of the bias pattern used to demonstrate the 1T-DRAM functionality of the partially depleted SOI MOSFET. In the bottom side, the driven current is monitored. Initially, the holes are injected in the floating-body by impact ionization due to the large current driven by the device (V_D=1.6V while V_G>V_T, W '1' in Fig. 2): the highly energetic electrons knock electrons out of their bound state and promote them to a state in the conduction band, creating electron-hole pairs. Electrons are evacuated through the drain, while holes are trapped into the neutral body of the silicon film. The hole overpopulation of the body of the device leads to a decrease in the threshold voltage and therefore a transitory increase of the drain current. The cell can be purged of charge by forward biasing the drain-to-body junction (negative drain bias, W '0' in Fig. 2).

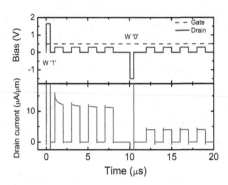

Fig. 2. Simulation results for the operation of a PD 1T-RAM based on the Z-RAM approach. The picture shows the bias pattern (top) and the driven current (bottom). For simplicity the gate bias is maintained constant ($V_G > V_T$). The floating body is initially charged with holes generated by impact ionization (W '1'), and the cell state is read four times by using a small drain bias. At t=10 μs, the cell is purged (write '0' state: W '0'), and read again four times reflecting the different with respect to the previous '1' states. L=1μm, T_{Si}=300nm, T_{ox}=3nm, T_{BOX}=400nm, N_A= 10^{17} cm^{-3}.

In this process, holes are evacuated from the floating-body through the channel-to-drain p-n junction. If the cell state is read again, the current level remains in the stable level (lower current). For simplicity, the gate bias has been maintained constant and larger than the threshold voltage ($V_G > V_T$) during the whole simulation period in order to have always a conductive channel.

For the Z-RAM approach, Innovative Silicon reported a current margin between states around 10μA, on devices with L=400nm and W=25μm, operated at V_G=0.8V and V_D=0.2V during current sensing [16].

3.   Second Generation: Fully Depleted FB 1T-DRAM

Despite the floating body is a particular characteristic of PD-SOI MOSFETs, the storage of holes inside the body of a transistor can also be achieved in Fully-Depleted SOI transistors (FD-SOI) by applying a negative bias to the front-gate or back-gate (BOX). The negative bias creates an electrostatic potential well where holes can be stored.

To illustrate this effect, we show in Fig. 3 simulations results corresponding to a relatively thin (film thickness T_{Si}=70nm, buried oxide thickness T_{BOX}=50nm) SOI transistor with (Fig. 3a) and without (Fig. 3b) negative substrate bias. Because of the undoped thin-body the unbiased device (V_B = 0) lacks from floating-body effect, and the holes generated by impact ionization recombine so quickly that the difference between states is not noticeable (Fig. 3a). Again, the gate bias has been maintained constant and larger than the threshold voltage ($V_G > V_T$) during the whole simulation period.

By contrast, if a negative back-gate bias is applied to the device (emulating a second gate with the substrate and buried oxide), a potential well is formed and allows the accumulation of holes (Fig. 3b). The potential well acts as a container for the majority carriers. Holes can be injected and removed in the same way as in Fig. 2. Note than in this case both '1' and '0' current levels become unstable: for the '1' state, the potential well becomes overpopulated with holes, and equilibrium is recovered by recombination (current overshoot); for the '0' state the back interface is driven into depletion, but holes are restored by junction leakage and thermal generation tending to corrupt the '0' state and convert it to '1' state. Purging the cell by forward biasing the drain-to-body junction leads an under-population of holes (current undershoot).

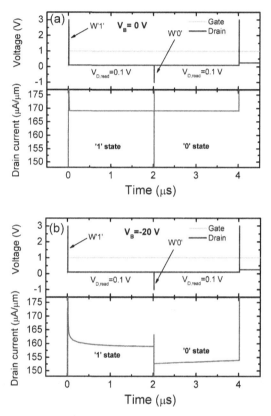

Fig. 3. Simulation results for the operation of a fully depleted FB 1T-RAM. The picture shows the bias pattern (top) and the driven current (bottom). For simplicity the front-gate bias is maintained constant ($V_G > V_T$). In Fig. (a) the substrate is biased at 0V, and due to the lack of holes in the floating body the memory effect is not manifested. In (b) a negative back-gate bias is applied creating a potential well where holes can be stored. The difference between the drain currents in '1' and '0' states represents the memory effect. L=200nm, T_{Si}=70nm, T_{ox}=1nm, N_A=10^{16} cm^{-3}. The writing pulses (W'1', W'0') appear as very short spikes in the large time scale.

One of the first examples (Fig. 4, [17]) of the feasibility of a fully depleted FB 1T-DRAM cell was experimentally demonstrated by Toshiba for 90 nm technology node in a 128 Mb memory array (numerical simulations were performed for devices beyond the 32 nm technology). The mechanisms proposed to write the '1' and '0' states were, respectively, impact ionization and junction forward bias (forcing the holes out of the transistor). To improve the retention time and the current sensing margin, the CMOS process included LDD, a relatively high channel doping (3x10^{17}cm^{-3}) and P-doped field plate. In addition, Co silicide and Cu wirings were used to reduce the parasitic source and drain resistances which usually degrade the current signal. The Co silicide source and drain are also attractive in order to reduce the parasitic disturbance (i.e. current leakage) between cell to cell. The '1' ('0')-state programming voltages were +1.5 V (-2.3 V) for the front gate and +2.2 V (-1.5 V) for the drain. The plate bias V_{GB} was fixed at -2.5 V. A ΔV_{TH} of 420 mV and a retention time of 70 ms were achieved for a temperature of 85°C. A similar 1T-DRAM with a thinner BOX was demonstrated by Intel [18].

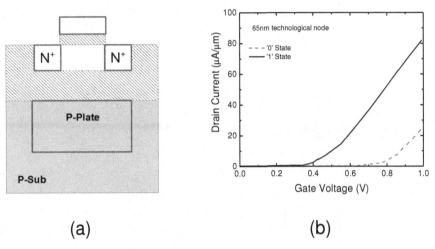

(a)   (b)

Fig. 4. Toshiba 1T-DRAM concept with implanted plate for back gate control [17]. (b) Typical current voltage characteristics in '0' and '1' memory states.

It has been shown that the devices exploiting FD technology present better performance, in terms of current margin between states and retention time, when the interface where the holes are stored is pushed into deep-depletion to create an unstable '0' state [19]. This 'capacitive coupling' mechanism is totally different from the case of the PD memory cells where the interface was overpopulated with holes to create an unstable '1' state. This memory cell is known as Meta-Stable-Dip RAM (MSDRAM) [19], and its working principle is based on the time-dependent subthreshold current in SOI MOSFETs [20]. An example based on a double-gated transistor is shown in Fig. 5. The reading condition of the sensing interface (back channel in Fig. 5) is set to a bias point slightly above threshold. In stable state, the front interface is accumulated ($V_{G1} < 0$) with holes, generated by band-to-band tunneling, and the MOSFET current flows through the opposite back interface. By contrast, if the storage interface is deep depleted (after ejecting the holes at $V_{G1} > 0$, the gate is pulsed back to -3 V), the potential in the body and at the sensing back interface decreases due to capacitive coupling, and the channel is cut. Retention time (governed by the parasitic repopulation of the front and back channels) is over 1s and several orders of magnitude in current margin between '0' and '1' states have been demonstrated in relatively large and thick transistors [19]. The MSDRAM shows attractive performance even in short devices if the film thickness is maintained above 25 nm [21,22].

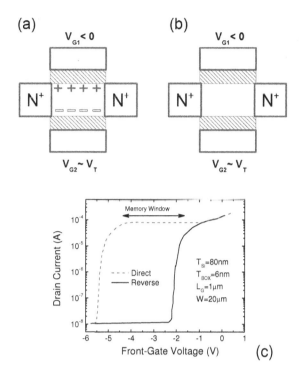

Fig. 5. Schematic representation of the double-gated memory cell MSDRAM proposed in [19]. (a) '1' state is stable leading to current flow at the back interface; (b) when the front interface is deeply depleted of holes, the potential at sensing back interface drops cutting the current flow. (c) Typical current voltage characteristics showing the hysteresis and the memory states in a MSDRAM.

A novel breaking approach was proposed by Okhonin in 2007 [23]. Instead of using the *pure* MOSFET operation to sense the cell state, it is possible to take advantage of the intrinsic bipolar transistor (N-P-N BJT) inherent to any MOSFET to read (and program) the memory cell. The N^+ source together with the P-type body and N^+ drain form the emitter, base and collector of the BJT, respectively (Fig. 6). The body of the MOS transistor (base of the bipolar transistor) is used as a storage node. To write a '1' into the memory cell, the intrinsic bipolar transistor is triggered, causing current to flow throughout the transistor body (Fig. 7a). This differs significantly from the MOS-like behavior where current flows only at the interface.

Charge collects at the interface due to the slight negative bias at the gate. The impact ionization effect used to create an excess of majority carriers in the floating body is more efficient in this bipolar bit cell structure, charging the body quickly and resulting in a very rapid write time. The state is read using a similar mechanism which senses the bipolar current through the transistor: when the top gate is populated with holes, retained by the negative gate bias, the bipolar transistor can be easily triggered by increasing the drain bias over the critical value to activate the bipolar conduction (V_{LATCH}, Fig. 7c). The large electron current during the bipolar action leads to a high impact ionization rate and hence a source of holes which feeds the base current of the bipolar transistor. The '0' state is written by removing the holes from the storage node (forward biasing the drain-to-body junction, Fig. 7b). When the gate is in lack of accumulated holes, the bipolar action cannot be triggered in the same reading bias conditions [24,25] (Fig. 7d). The current margin is relatively larger than in other FB-counterparts, however the large bias values (compared with other 1T-DRAM approaches for given device dimensions) required for the operation, introduces reliability, cell disturbance and power consumption issues.

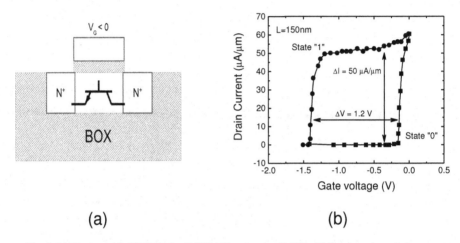

(a)   (b)

Fig. 6. (a) Bipolar mode 1T-DRAM cell (Z-RAM generation 2) [23]. (b) Programming window.

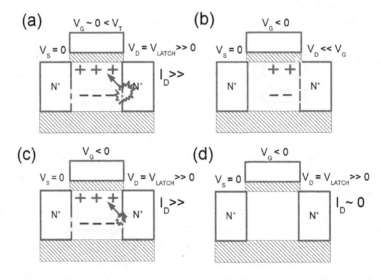

Fig. 7. Operation of a Z-RAM generation 2 cell. (a) Writing '1' by triggering the bipolar transistor. (b) Writing '0' by foward biasing the drain-body junction to remove the holes. (c) Reading '1' state: when the gate is accumulated with holes the bipolar action can be triggered. (d) Reading '0': when the gate is depleted of holes the bipolar transistor cannot be triggered.

This memory concept (denominated generation-2 of Z-RAM) was demonstrated in array using a standard 90nm SOI process. According to the experiments the current margin on 150nm gate-length devices was 50μA/μm, whereas the retention time at 85°C on L=220nm devices was always over 30ms [23].

The threshold voltage (V_T) feedback in SOI devices [8] was proposed in [26] as a new programming mode for FB 1T-DRAM. The bipolar action is not necessary for the occurrence of the floating-body hysteresis. The mechanism can be predominately driven by the V_T-feedback loop that works as follows (see Fig. 8). In the sub-V_T region, the hole current (I_{sub}), which is generated by impact ionization at the drain, can give rise to an increase of the body potential (V_{BS}). The positive V_{BS} has an exponential influence on the subthreshold current by decreasing V_T ($V_T(I_h)$ in Fig. 8); more holes will be generated and injected to the body to raise V_{BS} further. Once this feedback loop gain is larger than unity, positive feedback will occur. During opposite scan (decreasing V_G), the excess holes will temporarily prevent the current to drop and the MOSFET will latch.

Excess holes are injected into the body to write '1' state. Removing these injected holes either by forward-biasing the drain junction or by capacitive coupling results in a state '0'. Typically, the V_T-feedback loop is not observed because most of the devices are designed for logic applications and the body factor is ~0.3 [13], which is not large enough to result in positive feedback. With FDSOI, since V_T is insensitive to the V_{BS}, the V_T-feedback loop is of minor importance. But, UTBOX FDSOI (Ultra-Thin Buried Oxide Fully-Depleted SOI) offers the possibility to demonstrate this feedback. For the state reading, the back-channel current is sensed, i.e., the device is biased with negative front-gate bias V_{FG} and positive back-gate bias V_{BS}. The back channel does conduct when holes are present in the front interface, and does not conduct when holes are absent or swept away from the body. Retention time over 0.1s was reported in L=55nm devices [26].

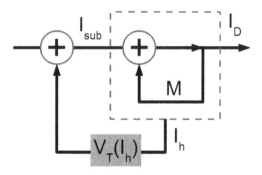

Fig. 8. V_T-feedback loop, M is the impact ionization multiplication factor.

The Z^2-FET is a recent device that shows attractive 1T-DRAM performance [27] based on field-effect-controlled charge regeneration. It is actually a forward-biased PIN diode, where the fully depleted body is partially covered by the gate (Fig. 9a). The front and back gates are biased such as to form potential barriers preventing the injection of electrons and holes from the N and P contacts, respectively. The biasing of the two gates emulates a thyristor configuration without needing any body doping. The current remains low until V_D increases enough to lower the electron injection barrier. A few electrons are injected from N contact into the channel and flow to the source where they reduce the hole injection barrier. The holes flow from source to drain, causing positive feedback that turns on the device and eliminates the injection barriers. This mechanism results in a strong $I_D(V_D)$ hysteresis (Fig. 9b) which is gate controlled and useful for capacitor-less memory. The states '1' or '0' are programmed by storing or not holes under the gate. Memory reading consists in discharging the gate. In '1' state, the discharge current is sufficient to turn on the Z^2-FET and the read current is high. In '0' state, there is no discharge current, hence the diode remains blocked (negligible read current). Very fast read pulse (1 nsec) enables to minimize the amount of stored charge requested to trigger the device. The memory is scalable down to 30 nm gate lengths and offers long retention, 1 V operating voltage, and regenerative (nondestructive) reading.

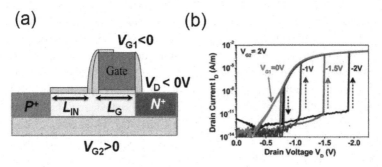

Fig. 9. (a) Configuration of the Z^2-FET memory and (b) typical $I_D(V_D)$ characteristics showing hysteresis controlled by the gate bias.

4. Electron-Hole Separation

One of the more questioned drawbacks of FB 1T-DRAM has been their scalability. As in conventional transistor + capacitor DRAM memory cells, the bit of information relies on the charge stored and consequently on the volume or area devoted to this storage. On the other hand, the scalability of Fully Depleted SOI transistors requires a decrease in the film thickness in order to suppress short-channel effects. In the case of single gate SOI technology, the film thickness of the device should be around four times thinner than the gate length (this condition can be relaxed by a factor of two or three in the case of double- or triple-gate devices) [13]. This means that FB 1T-DRAMs should be compatible with body thicknesses below 10nm in order to be competitive in future technology nodes. This condition imposes a capital challenge for the FB 1T-DRAM family. Several studies have shown severe degradation in the current margin between the memory states when the body thickness of 1T-DRAMs decreases below 30nm [28]. This limitation, also known as super-coupling effect [29], is basically an electrostatic consequence: the thinner the body, the more difficult to achieve the potential difference in order to accommodate a high concentration of electrons at one interface and a high concentration of holes at the other interface of the same silicon slab.

In order to overcome this intrinsic limitation (for maintaining the retention and sensing margin performances despite the scaling of the film thickness), several architectures and material combinations have been proposed. The idea behind all of them is the separation of the stored carriers and the sensing carriers by creating dedicated volumes (potential wells) inside the transistor body (multi-body devices).

The *single-transistor quantum well* (QW) 1T-DRAM [30], Fig. 10a, uses an engineered body integrating within the Si film a thin layer of a material with a narrower band gap (i.e. SiGe). This layer serves as storage well for holes (Fig. 10b). It was theoretically demonstrated that this structure improves the current sensing margin and scalability characteristics. Compared with FB 1T-DRAM, this QW memory has the ability to store the holes closer to the front-gate inducing an enhanced V_{TH} shift and retention time. Thanks to the introduction of the extra "storage room", QW devices are more scalable because the effect of the volume reduction with the channel length is lessened.

The performance of QW structures can be enhanced by using GaP raised Source/Drain (RSD) regions [31], implementable in either planar SOI or FinFET process flow. For the same gate length, the RSD structure has higher volume to store the charge inside the body. More importantly, GaP has a much higher bandgap than Si providing a better hole confinement within the body and hence improving the retention time.

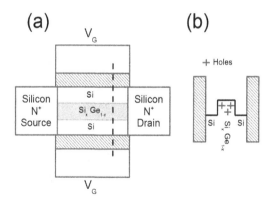

Fig. 10. Engineered body 1T-DRAM with a quantum well (a) and detail of the potential well created by the SiGe layer (b).

A *convex channel 1T-DRAM* structure (Fig. 11a) using the BJT programming technique was proposed to improve the retention time in [32]. The holes are stored beneath a raised gate oxide which may be filled by a smaller bandgap material (e.g. SiGe, Fig. 11b). As the holes stored during the 1-state programming reduce the body/source (drain) potential barrier, they easily diffuse through these junctions filling the SiGe region. The convex channel architecture provides a physical well for more effective storage of holes. Moreover, if a smaller bandgap material is used in the convex channel region, a deeper potential well is formed improving further the sensing margin and retention time.

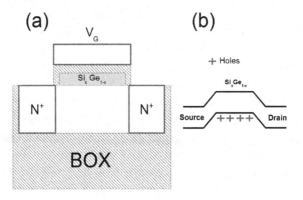

Fig. 11. (a) Convex channel structure filled with SiGe; (b) potential well created by the SiGe layer gate-stack.

Another alternative, named ARAM, was proposed in [33,34] by physically dividing the body of the transistor in two isolated regions (Fig. 12). In order to take advantage of this structure the Middle Oxide (MOX), which divides the body, must present a dielectric constant smaller than in silicon (for example SiO_2).

The resulting *semibodies* share the source and drain regions of the transistor. When this device is operated as a memory cell, the top semibody is used for majority carrier storage (holes) accommodated in a potential well created by the negative bias of the gate. The bottom semibody serves to sense the device state through a minority carrier current. The low-k MOX constitutes the key advantage of this device: the electrostatic potential difference between the front- and back-interfaces is enlarged due to larger potential drop through the MOX.

Memory state '1' (Fig. 12a) is programmed by charging the top semi-body with holes via impact ionization or band-to-band tunneling. When the semibodies thickness is below 10nm a volume inversion channel is activated by electrostatic coupling in the ultrathin bottom semi-body, establishing electrical continuity between source and drain regions. If the drain voltage is increased, a substantial current flows through the transistor. When no charge is stored in the top semibody (Fig. 12b) the electron concentration in the bottom semibody (sense channel) is extremely low leading to a negligible level of current. Numerical simulations have shown current margin between the two memory states over three orders of magnitude with retention time over 100ms in devices with 12nm film thickness [33].

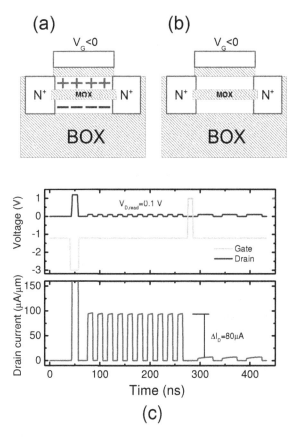

Fig. 12. Schematics of the FB 1T-DRAM (ARAM) proposed in [33]. (a) '1'state with charged semibodies. A high current is sensed by increasing the drain voltage; (b) '0' state with disgorged bodies. Increasing the drain bias leads to a low current (subthreshold level). (c) Simulation results for the operation of A-RAM cells. The bias scheme for memory programming, holding and reading is shown in the upper panel; the lower panel reproduces the corresponding current values.

5. Beyond SOI Substrates

FB 1T-DRAMs are intrinsically based on SOI technology. This has been pointed as an inconvenience, because SOI is not yet the leading CMOS technology. However, several attempts to export the SOI concepts to bulk substrates have been demonstrated experimentally or by numerical simulations.

In 2004, a FB 1T-DRAM cell was fabricated on bulk-Si substrate by emulating the SOI-like floating-body effect (Fig. 13). A triple-well NMOSFET was proposed with a buried n-implant [35]. The device was demonstrated on 0.13μm technology (Lg=0.35μm, W=0.3μm) with 3.5nm gate oxide. The programming of the '0' state was achieved by forward biasing the body-to-source junction whereas the '1' state was programmed by impact ionization. The initial current margin between states was 13μA, and the retention time was claimed to be 100ms at 25°C, with a limit detection condition of 1μA between states.

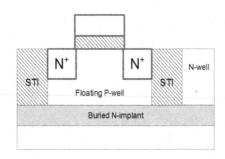

Fig. 13. FB 1T-DRAM on bulk substrate proposed in [34]. A deep N-implant is used to create the floating body.

Another breakthrough was the exportation of the second generation of Z-RAM technology to bulk silicon [36], without the requirement for SOI substrates, by using the 3D vertical transistor structures promoted by the major DRAM manufacturers (Fig. 14). With this variant, implemented in a test chip manufactured by Hynix Semiconductor, Innovative Silicon demonstrated that the Z-RAM could meet DRAM requirements for low power consumption, low-voltage operation, and the latest Double Data Rate (DDRx) performance levels. According to the experimental results, this memory cell, which uses the BJT programming method, provides one of the lowest operation voltage reported to date (drain voltage <1V). Array characterization demonstrated a programming window of 600mV (0% fail rate) and a retention of 900ms at 93°C.

Collaert's et co-workers have demonstrated a one-transistor capacitorless DRAM on standard bulk FinFET (Fig. 14b), using no additional processing. It was shown that, due to the use of the ground-plane doping and optimization of the READ bias conditions, no special process adjustment was required to obtain a programming window over 10μA and retention time over miliseconds was reported in L=130nm, W_{FIN}=20/30nm FinFETs [37].

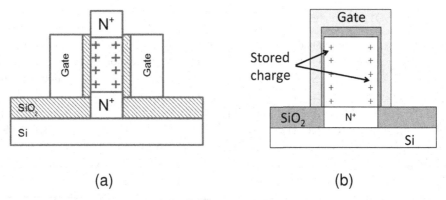

Fig. 14. Vertical double gate 1T-DRAM with common source [36]. (b) Bulk-FinFET based memory cell [37].

In 2011 an original concept of FB 1T-DRAM, maintaining the basic structure of a bulk MOSFET, device was presented [38]. The device, named A2RAM, is a multi-body single-gate device compatible with both SOI and bulk substrates without changes in its architecture. As shown in Fig. 15, the source and drain of a standard N-MOSFET are short-circuited by a N^+ film (*N-bridge*). The device is conceptually similar to the ARAM presented above (Fig. 12) except that there is no need for the middle oxide MOX to separate the two semi-bodies. The top P-body (below the gate) is lightly doped and becomes fully depleted due to the higher doping of the N-bridge. This lightly doped P-region stores the charge (holes) which defines the memory states whereas the N-bridge is used for current sensing. When the top P-body is populated with holes in accumulation (due to the negative gate bias), the gate-induced vertical electric field is screened by the charges (holes) and does not affect the carrier concentration (electrons) of the N-bridge. An electron current flows as soon as the drain bias is increased. By contrast, if the top P-body is temporarily depleted of holes, the gate field is no longer screened and fully depletes the N-bridge: the drain electron current becomes very low. The lack of holes in the top-body can be achieved by pulsing the gate bias to a positive voltage (which forward biases the source and drain junctions). When the gate bias returns to a negative voltage there is no immediate source to supply the holes. Fast restore of the hole population can be achieved by Band-To-Band (BTB) tunneling in the gate-drain region. The '1' state is stable (the holes accumulated by the negative gate bias can be maintained indefinitely). When the P-body is depleted, thermal generation, junction leakage, and Gate Induced Drain Leakage (GIDL) mechanisms tend to restore the hole population of the P-body, making the '0' state to degrade with time. Retention time over 100ms was predicted at 85°C in optimized 22nm-node SOI devices. The difference with conventional FB 1T-DRAM is three-fold: (i) the drain current (defining the cell state) is due to majority carriers flowing in the volume of the bridge, (ii) the use of an insulator substrate is optional, (iii) the supercoupling effect is eliminated by the PN junction separating the two types of carriers. Ultrathin devices (below 20nm) are realistic. Preliminary measurements have indicated the validity of the A2RAM concept.

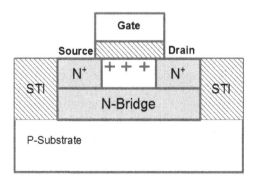

Fig. 15. Concept of A2RAM memory on bulk-Si wafers. The source and drain of a MOSFET are short circuited at a certain depth (~10 nm) by a high-doped N-type layer (N-bridge).

6. Conclusions

The floating-body 1T-DRAM family has grown exponentially during the last decade aiming to develop innovative memory cells and replace the standard 1T+1C DRAM which, after more than 30 years of unchallenged success, shows signs of inadaptability to further technology nodes. Each particular approach has its own advantages and drawbacks in terms of CMOS compatibility, scalability, power dissipation and performance. Some of them have stand up featuring promising characteristics, but no one is yet competitive enough to beat the DRAM supremacy.

There are several issues that FB 1T-DRAMs need to overcome before becoming a serious alternative:

(1) The use of the body of the transistor as storage node makes these cells prone to suffer from variability issues. In that sense, memory cells with negligible current in one of the states will be preferred.

(2) The retention time is mostly limited by carrier thermal generation mechanisms and BTB tunneling, which makes it difficult to achieve the standards of retention time of stand-alone DRAM in scaled devices. However, embedded memories which tolerate shorter retention time (a few microseconds) could benefit from the FB-1T-DRAMs concepts.

(3) The compatibility with the CMOS process of large circuits is essential. Also, new memory matrix needs to be developed demonstrating the immunity to cell disturbance during reading and writing events.

(4) Last but not least, FB-1T-DRAMs need to prove their scalability. Several solutions can fulfill the requirements to get rid of the previous issues, but memory cells with channel length below 30nm, overcoming all the constraints, are still to be demonstrated from an experimental point of view.

Acknowledgments

This work was supported in part by the Junta de Andalucia under Research Project TIC2010-6902, by the Spanish Government under Research Project TEC2011-28660 and by BIOTIC-UGR Campus of Excellence.

References

1. Sharma, A.K., "Advanced Semiconductor Memories, Arquitectures, Design and Applications", Wiley, 2003.
2. Dennard, R. "Field effect transistor memory", Patent U.S. Patent No. 3,387,286 (1968).
3. Suname, H., Kure, T., Hashimoto, N., Itoh, K., Toyabe, T., Asai, S., "A corrugated capacitor cell (CCC)". IEEE Trans. Electron Devices 1984; 31(6):746–53.
4. Kim, K., "Perspectives on giga-bit scaled DRAM technology generation." Microelectron Reliab 2000; 40(11):191-206.
5. Farid Nemati and James D. Plummer, "A Novel Thyristor-based SRAM Cell (T-RAM) for High-Speed, Low-Voltage, Giga-scale Memories" Electron Device Meeting, 1999 IEDM'99 Technical Digest, pp.283-289, 1999.

6. Chen, C.D., Matloubian, M., Sundaresan, R., Mao, B.Y., Wei, C.C., Pollack, G.P., "Single-transistor latch in SOI MOSFETs," *Electron Device Letters, IEEE*, vol.9, no.12, pp.636-638, Dec. 1988.

7. Dongwook, S., Fossum, J.G., "Dynamic floating-body instabilities in partially depleted SOI CMOS circuits," Electron Devices Meeting, 1994. IEDM'94. Technical Digest, International, vol.94, pp.661-664, 11-14 Dec 1994.

8. Ouisse, T., Ghibaudo, G., Brini, J., Cristoloveanu, S., Borel, G., "Investigation of floating body effects in silicon-on-insulator metal-oxide-semiconductor field-effect transistors" Journal of Applied Physics, 70, no.7, 3912-3919 (1991).

9. Wei, A., Sherony, M.J., Antoniadis, D.A., "Transient behavior of the kink effect in partially-depleted SOI MOSFET's," Electron Device Letters, IEEE, vol.16, no.11, pp.494-496, Nov 1995.

10. Tack, M.R., Gao, M., Claeys, C., Declerck, G.J., "The multistable charge-controlled memory effect in SOI MOS transistors at low temperatures," Electron Devices, IEEE Transactions on, vol.37, no.5, pp.1373-1382, May 1990.

11. Wann, H.J., Hu, C., "A capacitorless DRAM cell on SOI substrate," Electron Devices Meeting, 1993. IEDM'93. Technical Digest, International, vol.93, pp.635-638, Dec 1993.

12. Cristoloveanu, S., Li, S.S., "Electrical Characterization of Silicon On Insulator Materials and Devices." Kluwer Academic Publishers, Boston, ISBN 0-7923-9548-4, (1995).

13. Colinge, J.P., "Silicon-on-insulator technology: materials to VLSI", 3rd Ed, Klumer 2004.

14. Okhonin, S., Nagoga, M., Sallese, J., Fazan, P., "A SOI capacitor-less 1T-DRAM concept". In: Proceedings of the 2001 IEEE international SOI conference, Durango, USA; 2001. p.153-4.

15. Ioannou, D.E., Cristoloveanu, S., Mukherjee, M., Mazhari, B., "Characterization of carrier generation in enhancement-mode SOI MOSFET's." IEEE Electron Device Letters, EDL-11, no.9, 409-411 (1990).

16. Okhonin, S., Nagoga, M., Sallese, J.M., Fazan, P., "A capacitor-less 1T-DRAM cell," *Electron Device Letters, IEEE*, vol.23, no.2, pp.85-87, Feb 2002.

17. Hamamoto, T., Minami, Y., Shino, T., Kusunoki, N., Nakajima, H., Morikado, M., Yamada, T., Inoh, K., Sakamoto, A., Higashi, T., Fujita, K., Hatsuda, K., Ohsawa, T., Nitayama, A., "A floating-body cell fully compatible with 90-nm CMOS technology node for a 128-Mb SOI DRAM and its scalability". Electron Devices, IEEE Transactions on. 54: 563-571 (2007).

18. Ban, I., Avci, U.E., Shah, U., Barns, C.E., Kencke, D.L., Chang, P., "Floating Body Cell with Independently-Controlled Double Gates for High Density Memory", IEEE International Electron Device Meeting, pp.1-4 (2006).

19. Bawedin, M., Cristoloveanu, S., Yun, J.G., Flandre, D., "A new memory effect (MSD) in fully depleted SOI MOSFETs." Solid-State Electronics, 49, n9, 1547-1555 (2005).

20. Assaderaghi, F., Chen, J., Solomon, R., Chan, T., Ko, P., Hu, C., "Time Dependence of Fully Depleted SOI MOSFET's Subthreshold Current," Proc. IEEE International SOI Conf., Vail, Colorado, October 1991, pp. 32-33.

21. Bawedin, M., Cristoloveanu, S., Flandre, D., "A capacitor-less 1T-DRAM on SOI based on double gate operation." IEEE Electron Device Letts., 29, n7, 795-798 (2008).

22. Hubert, A., Bawedin, M., Guegan, G., Ernst, T., Faynot, O., Cristoloveanu, S., "SOI 1T-DRAM cells with variable channel length and thickness: Experimental comparison of programming mechanisms." Solid-State Electronics, 65–66C, November/December, 256-262 (2011).

23. Okhonin, S., Nagoga, M., Carman, E., Beffa, R., Faraoni, E., "New Generation of Z-RAM", EEE International Electron Devices Meeting, pp.925-928, 2007.

24. Giusi, G., Alam, M.A., Crupi, F., Pierro, S., "Bipolar Mode Operation and Scalability of Double-Gate Capacitorless 1T-DRAM Cells," *Electron Devices, IEEE Transactions on*, vol.57, no.8, pp.1743-1750, Aug. 2010.

25. Moon, D., Choi, S., Han, J., Kim, S., Choi, U., "Fin-Width Dependence of BJT-Based 1T-DRAM Implemented on FinFET," *Electron Device Letters, IEEE*, vol.31, no.9, pp.909-911, Sept. 2010.

26. Lu, Z., Collaert, N., Aoulaiche, M., De Wachter, B., De Keersgieter, A., Schwarzenbach, W., Bonnin, O., Bourdelle, K., Nguyen, B.Y., Mazure, C., Altimime, L., Jurczak, M., "A Novel Low-Voltage Biasing Scheme for Double Gate FBC Achieving 5s Retention and 10^{16} Endurance at 85°C", IEDM, pp.12.3.1-12.3.4, 2010.

27. Wan, J., Le Royer, C., Zaslavsky, A., Cristoloveanu, S., "A Compact Capacitor-Less High-Speed DRAM Using Field Effect-Controlled Charge Regeneration", IEEE Electron Device Letters, 33, 179-181 (2012).

28. Avci, U.E., Ban, I., Kencke, D.L., Chang, P.L.D., "Floating body cell (FBC) memory for 16-nm technology with low variation on thin silicon and 10-nm BOX," *SOI Conference, 2008. SOI. IEEE International*, pp.29-30, 6-9 Oct. 2008.

29. Eminente, S., Cristoloveanu, S., Clerc, R., Ohata, A., Ghibaubo, G., "Ultra-thin fully depleted SOI MOSFETs: special charge properties and coupling effects." Solid-State Electron 2007; 51(2):239-44.

30. Ertosun, M., Kapur, P., Saraswat, K., "A highly scalable capacitorless double gate quantum well single transistor DRAM: 1T-QW DRAM. Electron Device Letters, IEEE. 29: 1405-1407 (2008).

31. Pal, A., Nainani, A., Gupta, S., Saraswat, K., "Performance Improvement of One-Transistor DRAM by Band Engineering", IEEE Electron Device Letters", vol.99, 1-3 (2011).

32. Cho, M., Shin, C., Liu, T., "Convex channel design for improved capacitorless DRAM retention time." Simulation of Semiconductor Processes and Devices, 2009. SISPAD'09. International Conference on. pp.1-4 (2009).

33. Rodriguez, N., Gamiz, F., Cristoloveanu, S., "A-RAM Memory Cell: Concept and Operation", IEEE Electron Device Letters, vol.31, no.9, pp.972-974, 2010.

34. Rodriguez, N., Cristoloveanu, S., Gamiz, F., "Capacitor-less A-RAM SOI memory: Principles, scaling and expected performance", Solid-State Electronics, vol.59, 1, pp.44-50, 2011.

35. Ranica, R., Villaret, A., Malinge, P., Mazoyer, P., Lenoble, D., Candelier, P., Jacquet, F., Masson, P., Bouchakour, R., Fournel, R., Schoellkopf, J.P., Skotnicki, T., "A one transistor cell on bulk substrate (1T-Bulk) for low-cost and high density eDRAM," *VLSI Technology, 2004. Digest of Technical Papers. 2004 Symposium*, pp.128-129, 15-17 June 2004

36. Kim, J., Chung, S., Jang, T., Lee, S., Son, D., Chung, S., Hwang, S., Banna, S., Bhardwaj, S., Gupta, M., Kwon, J., Kim, D., Popov, G., Gopinath, V., Van Buskirk, M., Cho, S., Roh, J., Hong, S., Park, S., "Vertical double gate Z-RAM technology with remarkable low voltage operation for DRAM application," *VLSI Technology (VLSIT), 2010 Symposium*, pp.163-164, 15-17 June 2010.

37. Collaert, N., Aoulaiche, M., Rakowski, M., Redolfi, A., De Wachter, B., Van Houdt, J., Jurczak, M., "Optimizing the Readout Bias for the Capacitorless 1T Bulk FinFET RAM Cell", IEEE Electron Device Letters, 30, 1377-1379 (2009).

38. Rodriguez, N., Cristoloveanu, S., Gamiz, F., "Novel Capacitorless 1T-DRAM Cell for 22-nm Node Compatible With Bulk and SOI Substrates," IEEE Transactions on Electron Devices, vol.58, no.8, pp.2371-2377, Aug. 2011.

PLASMONIC-BASED DEVICES FOR OPTICAL COMMUNICATIONS

DJAFAR K. MYNBAEV* and VITALY SUKHARENKO†

New York City College of Technology of the City University of New York
*dmynbaev@citytech.cuny.edu
†vitaly.s@hotmail.com

To meet the demand of delivering ever-increasing Internet traffic, optical network must response by increasing its transmission capacity. Since transmission capacity of an individual fiber is still well exceed the capacity of transmitters (TXs) and receivers (RXs), wavelength-division multiplexing (WDM), in which many TXs and RXs at the transmitting ends of a fiber are used to send and receive many signals, becomes the necessary technology for increasing the transmission capacity of each link of an optical network. This trend, however, demands for increasing density not only the TXs and RXs, but all other components at the sending and receiving ends of communications links. As the number of wavelengths in WDM configuration getting greater, the number of all these components that must be placed on one board has to increase too; hence, the density of packaging comes to micro- and even nano-scale. The TXs and RXs are produced in arrays on a chip quite similar to production of VLSI electronic circuits. At that scale, traditional optical operations used today in an optical-communications technology, such as launching light into optical fiber from TXs and directing light from optical fiber into RXs, multiplexing and demultiplexing individual channels (wavelengths), and electro-optical (E/O) and opto-electrical (O/E) conversions become problems primarily because of the diffraction limit. The problems associated with the diffraction limit are particularly acute for optical interconnects. One of the possible solutions to all these—and some other—problems could be the use of plasmonics. In the last years, the optical-communications industry shows a great interest in developing this topic, as the growing number of publications and practical results can attest.

This paper consists of two parts. The first part reviews the current trends in application of plasmonics in optical communications and the second part discusses the theoretical foundation of the proposed WDM demultiplexer and offers the scheme of possible implementation of the device.

Keywords: Plasmonics; optical communications; wavelength-division demultiplexer.

Application of Plasmonics in Optical Communications

Plasmonics [1] is a branch of science and technology dealing with coupling of photons to free electron oscillations at the interface between a conductor and a dielectric. Though the concept of plasmons was introduced more than a century ago (G. Mie, 1908), only today practical realization of the plasmonics becomes feasible thanks to new nanofabrication technology and powerful simulation tools.

Plasmonics' main entity is surface plasmon polaritons (SPPs), which are two-dimensional electromagnetic waves that propagate between conductors (metals) and dielectrics. These surface waves are excited when light strikes the dielectric-metal interface; the energy of the photons is transferred to the metal and resonantly excites the oscillations of free electrons. The electrons' response results in the creation of dynamic charges on the metal's surface; these charges, in turn, produce waves called SPPs [1]. SPPs can be considered, in essence, secondary EM radiation obtained in response to incident light; the SPPs wavelength is much smaller than that of excitation or emitted photons; this is why we usually say that SPPs guide light confined in tiny—below the diffraction limit—surface.

From the optical-communications standpoint, the most important feature of plasmonic is that the SPPs can be seen as new optical information carrier that enables signal manipulation at the scale below diffraction limit. The fundamental challenge in using a plasmonic signal propagating through a metallic material is that it experiences huge resistive and radiation losses at optical frequency; as a result, SPPs can travel a distance of only few microns. In addition, controlling and manipulating this signal is not an easy task either because no applied technology is available yet for reliably working with such a truly nanoscale signal.

Plasmonics, however, has the potential to combine the best properties of both electronic and photonic worlds; in addition, plasmonics allows for reducing light manipulation from three to two dimensions. All these features might lead to new technology in creation of integrated photonic circuits. Figure 1 shows artist's impression a chip that combines the nanoscale dimension of modern electronic components and mature technology of their fabrication with speed of propagation and high bandwidth of optical transmission [2].

Plasmonic devices for optical communications can be classified as passive and active, as Fig. 2 shows.

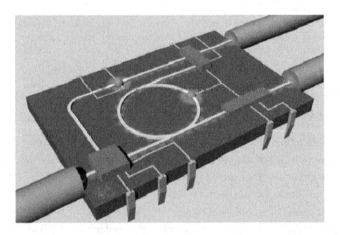

Fig. 1. Artist's view of a plasmonic chip [2]. (Reprinted with permission.)

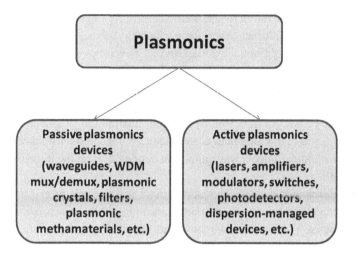

Fig. 2. Passive and active plasmonics.

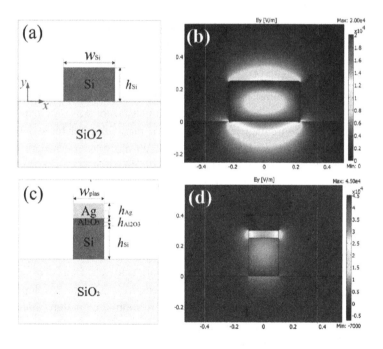

Fig. 3. Comparison of a silicon and plasmonic waveguides [3, 4]. (Reprinted with permission.)

Passive plasmonics refers to the area where SPPs—again, the bound electrons and photons oscillation at the metal surface excited by the incident light—travel along the interface without any additional external impact. Several potential plasmonic devices for optical communications are listed in Fig. 2. The plasmonic waveguide, as an example of passive plasmonics components, is shown in Fig. 3 [3, 4], where the top part shows standard silicon-on-isolator (SOI) waveguide and the bottom part demonstrates a plasmonic waveguide. Comparison of the field distribution in silicon [(a) and (b)] and plasmonic [(c) and (d)] waveguides clearly shows the much finer light confinement in the plasmonic waveguide. Such a waveguide can confine light at the dimensions of tenths of its wavelength.

Active plasmonics involves an external input that controls propagation and behavior of SPPs. In applications to optical communications, there are two main directions in developing active plasmonics: switching and modulation and amplification and lasing. One of the approaches to switching and modulation is based on the fact that being strongly confined inside the metal-dielectric interface SPPs are extremely sensitive to the change of refractive index of surrounding materials within a few tens of nanometers of this interface. Hence, changing the refraction index by light, thermal or electrical control signal will result in on-off switching and modulation of intensity and/or phase of propagating SPPs that would lead to creating all-optical switches and modulators at the subwavelength sizes. Figure 4 illustrates the concept of this approach where external excitation changes the refraction index either the metal or the dielectric component of an SPP waveguide [2]. This publication provides the comprehensive review of the current status of active plasmonics; in particular, it indicates that a metal-oxide-semiconductor (MOS) field-effect plasmonic modulator called plasmostor was reported. This plasmostor operates at gigahertz frequencies and tenths of picojoules-per-bit switching energy; its switching time still can't reach that of a silicon electro-optic switch and a transistor switch, but its switching energy is already smaller than that of silicon electro-optic switch though yet greater than that of a transistor switch.

In area of light amplification and lasing, it's worth to concentrate on such an impressive representative of active plasmonic components as a plasmon laser. Such a laser can generate and sustain light well below its diffraction limit. The main challenge in creating a plasmon laser is that the metallic cavity of a plasmon laser generally exhibit, as we mentioned above, high ohmic and radiation losses. Hence, to attain a sufficient gain, it was necessary to use cryogenic temperatures. Recent developments, however, brought new techniques leading to creating a room-temperature semiconductor sub-diffraction-limited laser. Radiation of such a laser is shown in Fig. 5 [5]. This laser confines light at about 20 nm, which is roughly one-twenty-fifth of its wavelength (the laser operates at wavelength around 500 nm) and is comparable to the size of a single virus.

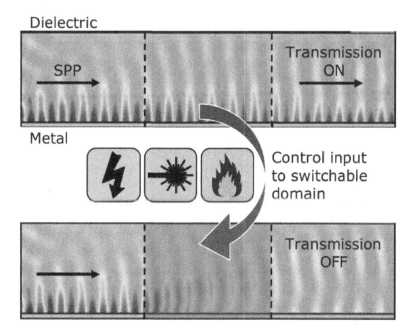

Fig. 4. A plasmonic switch as an example of active plasmonics [2]. (Reprinted with permission.)

Fig. 5. Radiation of plasmonic laser [6]. (Reprinted with permission from AAAS.)

WDM Demultiplexers

Among the plasmonic devices for optical communications, the great interest is hold in WDM demultiplexers for the following reasons: WDM beam emerging from an optical fiber contains today up to 80 wavelengths and each of these wavelengths must be directed to an individual photodiode (PD). On the other hand, to minimize the size and power consumption of a receiver, we usually use an array of PDs built in micro scale. Thus, the emerging beam with the diameter around 10 μm covers the whole array, which doesn't allow for working with individual PD. Figure 6 visualizes this problem. The other problem with today's WDM demultiplexers is that these devices are still bulky because it based on individual optical components. And there is another problem with the traditional demultiplexers: They must provide the maximum possible isolation between two adjacent channels (this parameter is also referred to as adjacent crosstalk).

Plasmonic WDM demultiplexer, by overcoming the diffraction limit and introducing the better characteristics in a smaller volume with minimal power consumption, might help in solving these problems. This is why the development of plasmonic demultiplexers caused the considerable interest in research community. Among many proposed plasmonic WDM demultiplexers, several schemes deserve mentioning here for introduction to this area of research.

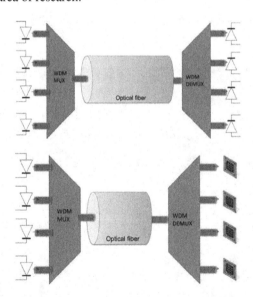

Fig. 6. Demultiplexing WDM signals with individual PDs (top) and PDs arrays (bottom).

Triple-wavelength tunable demultiplexer is shown in Fig. 7. This device uses two-dimensional (2D) metal-insulator-metal (MIM) waveguides and side coupled nanocavities (SCNCs). Incident light excites SPP waves on the both sides of the MIM structure. When properly designed rectangular nanocavity is placed on the sides of the MIM waveguide, incident light is partly coupled into the cavity. By changing the dimensions (lengths and widths) of the SCNCs, one can tune the wavelength passing the structure, as the bottom part of Fig. 7 shows. Hence, this design offers tunability, but it suffers with the limitation in the number of wavelengths that could be demultiplexed.

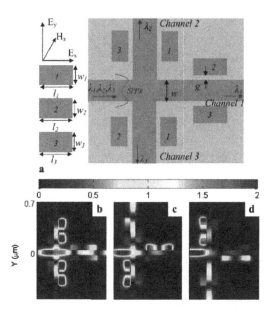

Fig. 7. Schematic diagram of the triple-wavelength WDM demultiplexer (top) and field distribution in the waveguides with incident wavelengths of 1310 nm, 1425 nm, and 1550 nm (bottom) [8]. (Reprinted with permission.)

A plasmonic dispersive demultiplexer shown in Fig. 8 is the other example of search of the most effective design [9]. The multiplexer is constructed from concentric gold-film groves with a radial width w, as shown in Fig. 8a. Point M is the center and R is the diameter of the focal circle. "The projections of the groves on the y axis are spaced equidistantly and form a grating with period d [9]." The groves are located on the grating arc with pole A. Figure 8b shows scanning electron micrograph (SEM) image of the groves for mode order 3. SPPs are excited in groves by the incident light and their diffraction depends on the mode order and the wavelength. The location of the focal point M changes along the focal circle depending of the SPP's wavelength and the mode order. Figure 9 demonstrates the SPP images for the first three modes and intensity profiles for combinations of two different wavelengths. Insets show the summation of SPP images around the focal areas; it's clearly seen two different foci for two different wavelengths.

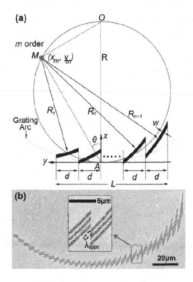

Fig. 8. Schematic diagram of plasmonic demultiplexer (a) an SEM image of the grating (b). Inset shows the details of a grove [9]. (Reprinted with permission.)

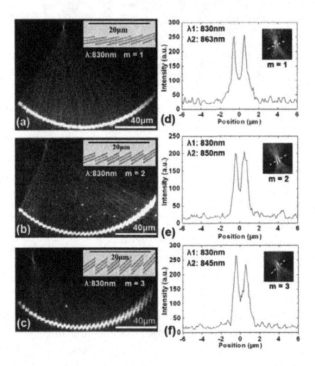

Fig. 9. Operation of the plasmonic demultiplexer [9]: Figures a), b) and c) show SPPs images of various modes (m = 1, 2 and 3); Figures d), e) and f) illustrates intensity profiles and focal areas (insets) of SPP images for three modes and various combinations of two wavelengths. (Reprinted with permission.)

A multiple-wavelength plasmonic demultiplexer is presented in paper [10] and shown in Fig. 10. The demultiplexer is fabricated on a glass substrate covered with metal film through which a nonperiodic rectangular slits array is formed. Each slit is about half-wavelength in size in *y* direction. The incident WDM beam strikes the slit array and excites multiple-wavelength SPPs; they propagate along the glass-metal interface. The key point here is the proper design of the slit pattern, which enables focusing the SPPs of different wavelengths at different spatial points, as shown in Fig. 10a. The spectra of measured photocurrent obtained by an integrated silicon PD and shown in Fig. 10b demonstrate that wavelengths of 820 nm, 850 nm and 880 nm are sorted at different spatial focal points. Interesting feature of this demultiplexer is that these focal points hops discretely from one position to the other as the wavelength is swept, whereas the conventional periodic grating devices exhibit continuous change of spatial positions as the wavelengths get swept. This design, as well as the previous one, features a good spectral and spatial resolution, but it works only with the wavelengths around 850 nm applicable for short-haul links with VCSELs, whereas the real need from WDM demultiplexing lies in long-distance networks operating at 1550 nm.

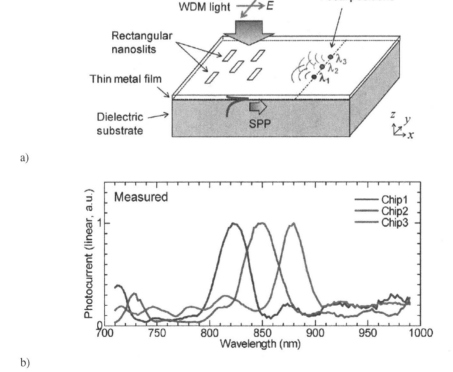

a)

b)

Fig. 10. Plasmonic demultiplexer with nonperiodic slit coupler: a) Schematic diagram; b) measured photocurrent [10]. (Reprinted with permission.) Copyright [2011] American Chemical Society.

A WDM demultiplexer involving the Airy plasmons was reported in [11]. Two-dimensional Airy beams are created by exciting SPPs at the surface of a metal film. These nondiffracting Airy plasmons can propagate along the curved trajectories, as was predicted by theory and confirmed by the experiment. Further theoretical studies showed that, using metal-dielectric-metal wedged structure with tilted metal plates (top figure), Airy plasmons generated at different wavelengths will be steered in different directions (bottom figure with false colors). This WDM demux might have a problem with harnessing light at the PDs active surfaces.

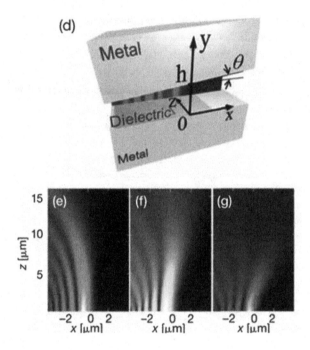

Fig. 11. Airy-plasmon WDM demultiplexer [12]: Schematic diagram (top) and artistic view of spatial distribution of different wavelength (bottom with false colors).

Proposed WDM Demultiplexer

In this section we will describe the proposed WDM demultiplexer; our theoretical part is based on the T-matrix method introduced by P.C. Waterman in 1969. Though our derivations based on the well-known theory, we use notations and some results introduced in [10]. We consider a two dimensional problem with $\frac{\partial}{\partial y} = 0$. Omitting time dependence, $\exp(j\omega t)$, we obtain wave equation in Cartesian coordinate system as follows:

$$\frac{\partial^2 \psi_i}{\partial x^2} + \frac{\partial^2 \psi_i}{\partial z^2} + k_i^2 \psi_i = 0 \tag{1.1}$$

where k_i is the wave number of the ith medium; i.e., $k_i = n_i 2\pi/\lambda = n_i \omega/c$, n is the refractive index, λ is the wavelength, ω is the radian frequency, c is the speed of light, and

$$\psi_i = \begin{cases} E_i - TE\ wave \\ H_i - TM\ wave \end{cases} \tag{1.2}$$

We consider the setup composed from three media, (1) glass with refractive index $n_g = 1.5$, (2) silver with refractive index $n_s = 0.1374 - 10.9512i$ at incident wavelength 1500nm, and (3) air with refractive index $n_a = 1$. Surfaces between each layers (1) and (2) and between (2) and (3) are described by functions S_0 and S_1, respectively. See Fig. 12. Incident plane wave strikes surface S_0 at angle θ_{inc} and secular reflection of the mode zero is reflected at θ_r. Thickness of the silver medium is d, grating pitch of the top surface is P_0 and grating pitch of the bottom P_1.

We use Greens functions as well as Huygens' Principle with appropriate boundary conditions. In order to describe the field in the first medium we use following equation:

$$\psi_{inc} - \int G_0 \frac{\partial}{\partial n_0} \psi_0 - \psi_0 \frac{\partial}{\partial n_0} G_0\ dS_0 = \begin{cases} \psi_0 & for\ z > S_0 \\ 0 & for\ z < S_0 \end{cases} \tag{1.3}$$

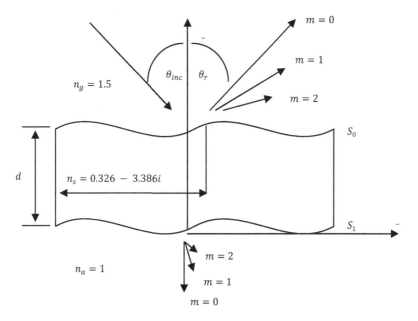

Fig. 12. Setup of the proposed demultiplexer.

where ψ_{inc} is the incident plane wave, ψ_0 is the field in the glass, n_0 is the normal with respect to the surface S_0, and G_0 is the Greens function for the first medium. We are using the Ewald - Oseen extinction theorem for $z < S_0$. To describe field in the middle medium, we use Greens function in the following form.

$$\int G_1 \frac{\partial}{\partial n_1} \psi_1 - \psi_1 \frac{\partial}{\partial n_1} G_1 \, dS_1 - \int G_1 \frac{\partial}{\partial n_0} \psi_1 - \psi_1 \frac{\partial}{\partial n_0} G_1 \, dS_0$$
$$= \begin{cases} 0 & for \; z > S_0 \\ \psi_1 \\ 0 & for \; z < S_1 \end{cases} \tag{1.4}$$

where ψ_1 is the field in the second medium, n_1 is the normal with respect to the surface S_1, and G_1 is the Greens function for the second medium. And for the third medium we obtain

$$\int G_2 \frac{\partial}{\partial n_1} \psi_2 - \psi_2 \frac{\partial}{\partial n_1} G_2 \, dS_1 = \begin{cases} 0 & for \; z > S_1 \\ \psi_2 & for \; z < S_1 \end{cases} \tag{1.5}$$

where ψ_2 is the field in the air and G_2 is the Greens function for the third medium. Then we apply boundary conditions,

$$\psi_0 = \psi_1 \; on \; the \; S_0$$
$$\psi_1 = \psi_2 \; on \; the \; S_1$$
$$\gamma_1 \frac{\partial}{\partial n_0} \psi_0 = \frac{\partial}{\partial n_0} \psi_1 \; on \; the \; S_0$$
$$\gamma_2 \frac{\partial}{\partial n_1} \psi_1 = \frac{\partial}{\partial n_1} \psi_2 \; on \; the \; S_1 \tag{1.6}$$

where $\gamma_1 = \frac{\varepsilon_2}{\varepsilon_1}$ and $\gamma_2 = \frac{\varepsilon_3}{\varepsilon_2}$ are the appropriate boundary condition coefficients for transverse magnetic (TM) waves.

The wave number can be decomposed into k_x and k_z propagating in the direction x and z respectively; thus, using diffraction equation,

$$k_{xm} = k_0 \sin \theta_{inc} + m \frac{2\pi}{P_0}, \tag{1.7}$$

we find:

$$k_{izm} = \begin{cases} \sqrt{k_i^2 - k_{xm}^2} & for \; \dfrac{k_{xm}}{k_i} \geq 1 \\ -j\sqrt{k_{xm}^2 - k_i^2} & for \; \dfrac{k_{xm}}{k_i} \leq 1 \end{cases} \tag{1.8}$$

To make our derivations more explicit, we use some our results obtained previously [13]. Using simple manipulation and boundary conditions we can find $\varepsilon_1 k_{0zm} = \varepsilon_0 k_{1zm}$ at the surface S_0 and $\varepsilon_2 k_{1zm} = \varepsilon_1 k_{2zm}$ at the surface S_1, where ε_i is the electric permittivity of the medium. Now we can derive

$$\varepsilon_1 \sqrt{\frac{\omega^2 \varepsilon_0}{c^2} - k_{xm}^2} = \varepsilon_0 \sqrt{\frac{\omega^2 \varepsilon_1}{c^2} - k_{xm}^2} \tag{1.9}$$

$$\varepsilon_1^2 \left(k_{xm}^2 - \frac{\omega^2 \varepsilon_0}{c^2} \right) = \varepsilon_0^2 \left(k_{xm}^2 - \frac{\omega^2 \varepsilon_1}{c^2} \right). \tag{2.1}$$

Solving for the k_{xm}^2, we find

$$\frac{\omega^2}{c^2} \frac{\varepsilon_1 \varepsilon_2}{(\varepsilon_1 + \varepsilon_2)} = k_{xm}^2 \tag{2.2}$$

Refer to Fig. 13 where we plot k_{xm} vs. incident ω, both normalized with respect to plasmonic frequency, $\omega_p^2 = \frac{ne^2}{\varepsilon_0 m}$, where n is number density of free electrons, m is an effective optical mass of an electron, and e is the charge of an electron. Pay attention to the straight lines that represent light lines of air and silica. Bound nature of SPPs excitation corresponds to the dispersion curves located higher of the light lines. Consider dispersion curve of silica (solid curve): When ω/ω_p is low ($\omega/\omega_p = 0.3$, for example, which corresponds to low- and mid-infrared range), the dispersion curve lies very close to the light line and SPPs acquire properties of Sommerfield-Zeneck waves; in other words, when the wave number is small, SPPs exhibit as weak near fields. On the other hand, when ω/ω_p is high (consider $\omega/\omega_p = 0.5$), which means that the wave number becomes large, the SPPs becomes much enhanced Surface Plasmon Polariton waves. Thus, k_{xm} is the measure of SPPs' strength. This phenomenon explains our choice for the setup consisting from silica, silver and air (Fig. 12): At the silica-silver (top) interface k_{xm} is small

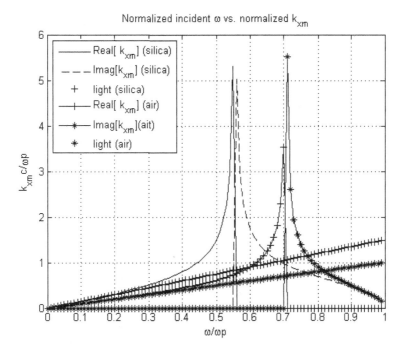

Fig. 13. Normalized wave number k_{xm} vs. normalized incident frequency.

whereas at the silver-air interface k_{xm} is large and SPPs are excited. In addition if the top grating pitch is decreased with respect to the bottom one, the difference in propagation vector in x direction will increase thus increasing window of SPPs excitation for higher modes, excluding zero mode.

Returning back to the mainstream of our derivations, we express fields ψ_i and Greens functions G_i in Fourier series and arrive to the set of Equations 2.3-2.7. (The fields are unknown, so by expressing them in Fourier series we limit our unknowns to the coefficients of the fields; these fields are not derived but assumed.)

$$\psi_0 = 2 \sum_n \alpha_{0n}^s \exp(-ik_{xn}x'), \qquad for\ z' = S_0 \qquad (2.3)$$

$$\frac{\partial \psi_0}{\partial n_0} = -2ik_0 dx' \sum_n \beta_{0n}^s \exp(-ik_{xn}x'), \qquad for\ z' \qquad (2.4)$$

$$\psi_{inc} = \sum_m \frac{a_m \sqrt{k_0} \exp(-i(k_{xm}x - k_{0zm}z))}{\sqrt{k_{0zm}}}, \qquad for\ z < S_0 \qquad (2.5)$$

$$\psi^r = \sum_m \frac{b_m \sqrt{k_0} \exp(-i(k_{xm}x + k_{0zm}z))}{\sqrt{k_{0zm}}}, \qquad for\ z > S_0 \qquad (2.6)$$

$$G_0 = -\frac{i}{2k_0 P} \sum_m \frac{k_0}{k_{0zm}} \exp(-i(k_{xm}(x - x') + k_{izm}|z - z'|)) \qquad (2.7)$$

where x' and z' are the surface coordinates, α_{0n}^s are the coefficients of the surface fields ψ_0, β_{0n}^s are the coefficients of the surface fields $\frac{\partial \psi_0}{\partial n_0}$, $a_0 = 1$ and $a_{m+1} = 0$ since we consider incident plane wave, and finally b_m are the coefficients of the scattered modes. Using appropriate boundary conditions we can equate surface coefficients to the incident and scattered coefficients.

$$\begin{bmatrix} b_m \exp(ik_{0zm}d) \\ a_m \exp(-ik_{0zm}d) \end{bmatrix} = \begin{bmatrix} -Q_D^+ & -Q_N^+ \\ Q_D^- & Q_N^- \end{bmatrix} \begin{bmatrix} \alpha_{0n}^s \\ \beta_{0n}^s \end{bmatrix} \qquad (2.8)$$

$$X = \begin{bmatrix} -Q_D^+ & -Q_N^+ \\ Q_D^- & Q_N^- \end{bmatrix}, \qquad (2.9)$$

where $Q_D^\pm$ and $Q_N^\pm$ are

$$Q_D^\pm = -\frac{\sqrt{k_0}}{P\sqrt{k_{0zm}}} \int_{-P/2}^{P/2} \exp(\pm ik_{0zm}S_0 - ix'(k_{xn} - k_{xm}))\, dx'$$

$$Q_N^\pm = \frac{k_0 - k_{xm}k_{xn}}{\pm k_{0zm}} Q_D^\pm \qquad (3.0)$$

Following the same procedure we relate coefficients of surface S_1 to the transmitted coefficients and obtain matrix Y similarly how we find the matrix X.

To derive the equation of the coefficients for the middle layer, we introduce the R-matrix,

$$\begin{bmatrix} \alpha_{0n}^s \\ \alpha_{1n}^s \end{bmatrix} = \begin{bmatrix} R_{11} & R_{12} \\ R_{21} & R_{22} \end{bmatrix} \begin{bmatrix} \beta_{0n}^s \\ \beta_{1n}^s \end{bmatrix} \tag{3.1}$$

where α_{1n}^s is the coefficient of the surface fields ψ_1 and β_{1n}^s is the coefficient of the surface fields $\frac{\partial \psi_1}{\partial n_1}$. Finally assembling all components together in one matrix, we can derive our T matrix solution as

$$\begin{bmatrix} b_m \exp{(jk_{0zm}d)} \\ A_m \end{bmatrix} = T \begin{bmatrix} a_m \exp{(-jk_{0zm}d)} \\ B_m \end{bmatrix} \tag{3.2}$$

with

$$T = \begin{bmatrix} X_{11}R_{11} + X_{12} & X_{11}R_{12} \\ Y_{21}R_{21} & Y_{21}R_{22} + Y_{22} \end{bmatrix} \begin{bmatrix} X_{21}R_{11} + X_{22} & X_{21}R_{12} \\ Y_{11}R_{21} & Y_{11}R_{22} + Y_{12} \end{bmatrix}^{-1} \tag{3.3}$$

where A_m are the modes of transmitted wave, and $B_m \equiv 0$ to satisfy Somerfield radiation conditions that state there is no incident light on the bottom surface.

Equations 3.2 and 3.3 are our main results that allow for describing the reflection and transmission in our setup.

We can extend this approach to the following problem: Calculations show that the transmitted modes A_m are concentrated in the near-field region, therefore it might be beneficial to add another layer between the silver-air interface to move the collection of the transmitted modes to the far-field region. R-matrix can be repeated to implement this scenario in the following way: First, we decompose previously calculated R-matrix in to A and B matrices and manipulate S_0 and S_1 surface coefficients from equation 3.1 to collect S_0 coefficients on the left hand side of the equation and S_1 coefficients on the right:

$$\begin{bmatrix} \alpha_{0n}^s \\ \beta_{0n}^s \end{bmatrix} = [A^{-1}B] \begin{bmatrix} \alpha_{1n}^s \\ \beta_{1n}^s \end{bmatrix}. \tag{3.4}$$

Similar process can be done to acquire the following equation for the coefficients at S_1 and S_2

$$\begin{bmatrix} \alpha_{1n}^s \\ \beta_{1n}^s \end{bmatrix} = [C^{-1}D] \begin{bmatrix} \alpha_{2n}^s \\ \beta_{2n}^s \end{bmatrix}. \tag{3.5}$$

It is quite clear at this point that matrices A, B, C and D can be combined to relay S_0 coefficients to S_2 coefficients:

$$\begin{bmatrix} \alpha_{0n}^s \\ \beta_{0n}^s \end{bmatrix} = [A^{-1}B \, C^{-1}D] \begin{bmatrix} \alpha_{2n}^s \\ \beta_{2n}^s \end{bmatrix} \tag{3.6}$$

We define new matrix as E,

$$E = B\, C^{-1}D, \tag{3.7}$$

and finally, using Equations 3.4 and 3.5, we can solve for the R-matrix

$$\begin{bmatrix} \alpha_{0n}^s \\ \alpha_{2n}^s \end{bmatrix} = \begin{bmatrix} R_{11} & R_{12} \\ R_{21} & R_{22} \end{bmatrix} \begin{bmatrix} \beta_{0n}^s \\ \beta_{2n}^s \end{bmatrix}, \tag{3.8}$$

where

$$[R] = \begin{bmatrix} A_{11} & -E_{11} \\ A_{21} & -E_{21} \end{bmatrix} \begin{bmatrix} -A_{12} & E_{12} \\ -A_{22} & E_{22} \end{bmatrix} \tag{3.9}$$

Now we can simply combine matrix of the first surface, R-matrix, and last surface matrix to obtain final solution, T-matrix, as in Equation 3.3.

Using previously described T-matrix approximation resulted in Equation 3.2, we calculated SPPs excitation for the grating with top grating pitch equals 15 μm and bottom grating pitch equals 0.15 μm at the angle around the total internal reflection. (Making the grating pitches so different we significantly increase effectiveness of SPPs excitation.) Figure 14 shows that reflection of the incident light drops to zero at the angle of the total internal reflection, indicating that all incident energy is transferred into the silver film and excites the SPPs. Note that transmission coefficients indicate the strength of SPPs and that is the reason for third order coefficients. The longer the wavelength, the stronger the SPPs, as demonstrated in Fig. 15. Therefore, longer wavelengths penetrate silver medium farther and excite stronger field of SPPs, as Fig. 16 shows.

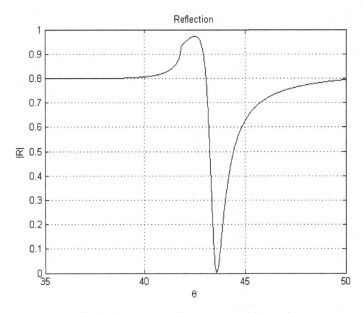

Fig. 14. Reflection coefficient vs. the incident angle.

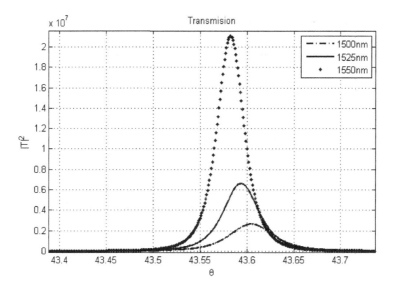

Fig. 15. Transmission coefficients vs. the incident angle.

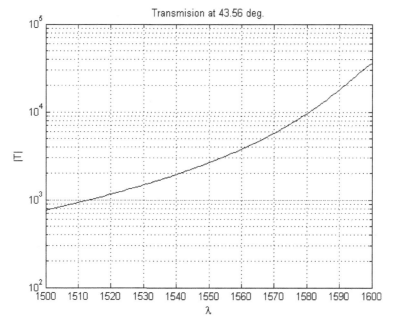

Fig. 16. Transmission coefficient vs. wavelength.

Figure 17 shows that the wave number in z direction increases with the increase of the wavelength, in total compliance with the abovementioned results. (Refer to Fig. 12 for x- and z-directions.) One can readily see that at the distance of 1 μm from silver-air interface, the intensity of 1525-nm signal will be four times less than that of 1550 nm and intensity of 1500-nm signal will be even eight times smaller than that of 1500 nm. This fact allows us to discriminate the signals based on their wavelengths by varying the placement of a PD along z axis. What's more, since the SPPs of different wavelengths decay differently along the silver-air interface, as Fig. 18 shows, there is the second degree of freedom in discrimination of signals of various wavelengths. For example, placing a PD at 3 μm from a starting point along x axis, we will practically collect only 1550-nm signal because 1500-nm and 1525-nm signals will diminish.

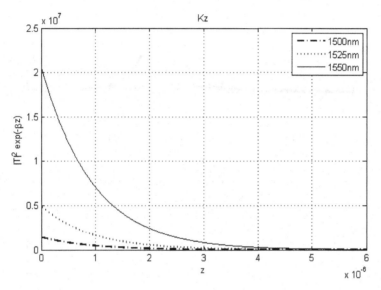

Fig. 17. Change of wave number k_z with distance z for various wavelengths.

Thus, our results lead to the following construction of the proposed plasmonic WDM demultiplexer: The special array of PDs is placed at the silver-air interface. Each PD is located in its individual position along both positive x axis and negative z axis, as Fig. 18 shows. In practical implementation, the displacement along z axis can be replaced by the proper attenuator with predefined attenuation for each PD. In the example of the setup shown in Fig. 18, the first PD collects all four wavelengths; the second PD collects $\lambda 2$, $\lambda 3$, and $\lambda 4$; the third PD collects $\lambda 3$, and $\lambda 4$; and the last PD collects only $\lambda 4$. A signal from the fourth PD delivered by $\lambda 4$ is fed into the output of the third PD, which allows for eliminating this signal from the third PD's output. Thus, the processed output of the third PD will contain only signal delivered by $\lambda 3$. Similarly, the outputs of the second and the first PDs will contain only signals delivered by $\lambda 2$ and $\lambda 1$, respectively. Finding the specific characteristics of the proposed plasmonic WDM demultiplexer is the task for the future work.

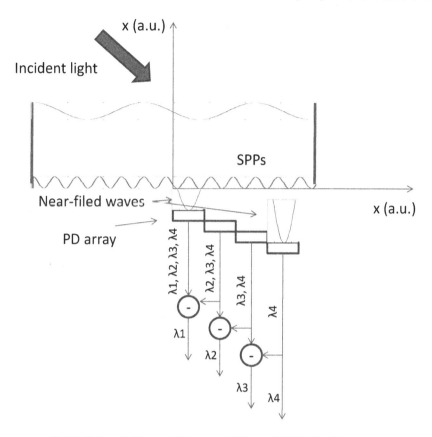

Fig. 18. Schematic diagram of the proposed plasmonic WDM demultiplexer.

Conclusion

We review the current status of application of plasmonics to optical communications and demonstrated a significant interest of the research community in developing plasmonic WDM demultiplexers. We propose the new scheme of the plasmonic demultiplexer, develop the theoretical tools for calculation of its major optical characteristics, perform the simulations that confirm our expectations and describe the operation of the proposed demultiplexer.

Acknowledgement

This work is partly supported by PSC-CUNY grant #63835-00 41.

References

1. Stefan A. Maier, *Plasmonics: Fundamentals and Applications*, Springer Science+Business Media LLC, 2007; Dror Sarid and William Challener, *Modern Introduction to Surface Plasmons*, Cambridge University Press, 2010.

2. Kevin F. MacDonald and Nikolay I. Zheludev, "Active plasmonics: current status," *Laser Photonics Rev. 4, No. 4, 562-567 (2010).*

3. Yi Song, "Plasmonic waveguides and resonators for optical communications applications," *Doctoral thesis in Microelectronics and Applied Physics, KTH School of Information and Communication Technology, Stockholm, Sweden, June 2011.*

4. Yi Song et al, "Broadband coupler between silicon waveguide and hybrid plasmonic waveguide," *Optics Express*, 19-12-13173 (2010).

5. Ren-Min Ma et al, "Room-temperature sub-diffraction-limited plasmon laser by total internal reflection," *Nature Materials Letters 10, Feb. 2010, 110-113.*

6. Volker J. Sorger and Xiang Zhang, "Spotlight on Plasmon Lasers," *Science, 333, 709-710 (2011).*

7. Harry A. Atwater et al, Novel "Plasmonic Devices for Nanophotonic Networks," *www.plasmonmuri.caltech.edu/news/PosterAFOSR.pdf (2011).*

8. H. Lu et al, "Nanoplasmonic triple-wavelength demultiplexers in two-dimensional metallic waveguides," *App. Phys. B (2011) 103:877-881.*

9. Chenglong Zhao and Jiasen Zhang, "Plasmonic Demultiplexer and Guiding," *ACS Nano, 4, 11, 6433-6438 (2010).*

10. Takuo Tanemura et al, "Multiple-Wavelength Focusing of Surface Plasmons with a Nonperiodic Nanoslit Coupler," *Nano Lett.* 2011, 2693-2698.

11. A. Minovich et al, "Airy Plasmons: Bending Light on a Chip," *Optics and Photonic News,* December 2011, P.35.

12. M. Ohki et al, "T-matrix Analysis of electromagnetic wave diffraction from a dielectric coated Fourier grating," *Progress In Electromagnetics Research*, PIER 53, 91-108, 2005.

13. Djafar K. Mynbaev and Vitaly Sukharenko, "WDM Demultiplexing by Using Surface Plasmon Polaritons," *International Journal of High Speed Electronics and Systems*, Vol. 20, No. 1, March 2011, Pp. 51-61.

SPINTRONIC DEVICES AND CIRCUITS FOR LOW-VOLTAGE LOGIC

DANIEL H. MORRIS*, DAVID M. BROMBERG, JIAN-GANG (JIMMY) ZHU† and LARRY PILEGGI‡

Department of Electrical and Computer Engineering, Carnegie Mellon University, Pittsburgh, PA
**danielmo@andrew.cmu.edu*
†*jzhu@ece.cmu.edu*
‡*pileggi@ece.cmu.edu*

This paper describes the design of digital logic circuits composed exclusively from magnetic devices. The logic level of a signal is embedded in the direction of steered currents, not voltages. The currents are steered by small (e.g., 2-3x) resistance changes. Sub-100 mV pulsed voltages power and synchronize the circuits. Logic gates are non-volatile, allowing for fully-pipelined logic that can achieve ultra-low energy for design examples.

Keywords: Spintronics; magnetoelectronics; magnetic logic; spin-transfer torque; magnetic tunnel junction.

1. Introduction

In recent years, the concept of moving beyond charge-based electronics by exploiting the spin polarization of electrons – spintronics – has attracted attention. These magnetoelectronic circuits exhibit non-volatility and offer the potential to operate at lower power than today's electronics. We first discuss several magnetic logic designs and then introduce our new approach, mLogic. Switching is driven by low energy spin-transfer torque, not external fields. Devices and structures are similar to existing MRAM devices. Circuits have the capacity for fanout and non-magnetic signal propagation while having CMOS-independent fabrication and operation. mLogic devices and circuits enable ultra-low voltage, magnetic logic. Furthermore, the non-volatile state enables new types of logic architectures. A highly efficient and novel FFT datapath implementation is described to illustrate the promising energy-efficiency that may be achieved with mLogic computational block designs.

2. Magnetic Logic Background

A number of proposals have been made to use magnetic devices to build logic circuits. One class of approaches uses field-based switching of magnetic state. A second class of approaches uses spin-polarized current-based switching. The latter uses MRAM devices but requires tight integration with CMOS. Fanout and global propagation of signals are possible, but process costs are higher and switching energy is increased compared to magnetic-only schemes. Here, we discuss some of these approaches to understand the breadth of concepts considered.

2.1. *Magnetic Logic Schemes with External Field-Driven Switching*

2.1.1. *Magnetic Quantum Cellular Automata*

An approach to magnetic logic known as magnetic quantum cellular automata (MQCA) [1] is founded on the principle of using coupled nanomagnets to perform logic functions. The placement of these nanomagnets relative to one another defines the logic function; there is no physical wiring between elements or gates. Dipole coupling between these magnets induces the magnetization in neighboring structures to align parallel or antiparallel with the magnetization of the "input" magnet, depending upon whether these "output" magnets are vertically- or horizontally-collinear (Figure 1a).

Porod et al. have developed a variety of structures based on this approach. A fabricated three-input majority logic gate [2] of these coupled nanomagnets demonstrated correct logic operation roughly 25% of the time, with an external field used to write the initial state to a set of input magnets. This result highlights one of the key challenges of the MQCA approach. For the coupling mechanism to work reliably, the nanomagnets must be nominally identical in shape, magnetic characteristics, and relative separation to one another. This is no easy task, and the authors of [2] acknowledged the lack of reliability was most likely a result of variation due to fabrication and field application.

2.1.2. *Domain Wall Logic*

Another approach to purely magnetic logic was introduced by Allwood et al. based on propagating domain walls along curved permalloy tracks [3]. A rotating external field is applied to push the walls around the corners of the four primary shapes required to build logic circuits: an inverter, AND gate, fanout structure, and crossover structure (Figure 1b). In the inverter, for example, a domain wall is pushed around a cusp in the track, which results in a 180° rotation of the magnetization from input to output. Shift registers and various logic gates have been demonstrated using this scheme.

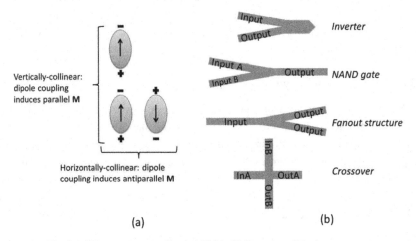

Fig. 1. (a) Nanomagnet coupling in MQCA. (b) Domain wall logic structures.

Although this approach is elegant in its simplicity, with no integrated switching elements, it does have its drawbacks. In particular, the on-chip generation of rotating magnetic fields to drive the domain walls is a challenge, especially if multiple field sources are needed. In this approach and in MQCA, high currents are required for field generation and the clock speed is unlikely to exceed several hundred MHz [4]. Perhaps most challenging, however, are the tight patterning constraints of these shapes. The tracks have a large area footprint (the domain walls themselves are wide), and it is unclear if these patterned arrangements of shapes will work as effectively at smaller feature sizes.

2.2. *Magnetic Devices with Spin-Transfer Torque Switching*

2.2.1. *Spin Transfer Torque MRAM*

The approaches to magnetic logic examined so far used magnetic fields to drive device switching. An alternative switching approach is based on spin-transfer torque (STT), where a spin-polarized electric current can directly program the magnetization of a material.

When an electron passes through a magnetized material (Figure 2), its spin becomes polarized in the direction of the magnetization (e.g., "spin-down" or "spin-up"). Once this spin-polarized electron reaches a region of magnetization of different orientation, its spin is again polarized to align with the magnetization. The spin re-alignment results in a change in the angular momentum of the electron. To conserve angular momentum, a torque is exerted on the local magnetic moment of the material. This *spin-transfer torque* acts on the magnetization to align it with the electron's initial polarization. For example, if a spin-down electron enters a region of magnetization oriented up, the spin-transfer torque will then act to pull that magnetization downward. This phenomena has been used to drive switching in both magnetic memories with domain walls [5][6] and single domain pillar MRAM [7][8]. Furthermore, STT has also been recognized to be useful for logic implementations [9].

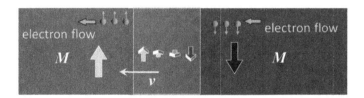

Fig. 2. Spin-Transfer Torque (STT) is basis for current-driven magnetic switching

A common way of sensing magnetization is with a magnetic tunnel junction (MTJ). A fixed magnetic electrode and switchable magnetic electrode sandwich a tunnel barrier. Flipping the magnetization of the switchable magnetic electrode switches the resistance through the barrier from its lowest stable value (R_L) to its highest stable value (R_H), or vice versa. This switching resistance phenomenon is known as tunnel magnetoresistance. The resistance is at its minimum value when the moments on either side of the barrier are parallel and maximum when the moments are antiparallel. Given today's technology, the resistance can change by a factor of 2-3x (i.e. 100%-200% TMR) [10][11].

2.2.2. *MTJ-Based Non-Volatile Logic Circuits*

The majority of today's MRAM devices are variants of STT-MTJs. In several proposed non-volatile logic gates, STT-MTJs are interconnected with CMOS to implement a logic function. CMOS circuits detect and amplify small resistance changes in a STT-MTJ devices [12]. The sense amplifier-like circuits then can drive fanout, switching the STT-MTJs in the next logic stage. These schemes that use variants of MRAM cells require tight integration with CMOS and raise the cost of fabrication compared to a non-heterogeneous technology. Furthermore, powering CMOS in these hybrid circuits prevents the low operating voltages possible with magnetoelectronics alone. It is recognized that many existing approaches to magnetoelectronics may have worse energy-delay products than CMOS once the circuit overhead is considered [13].

3. New Approach: STT-MTJ mCell

The proposed mLogic is based on a novel magnetoelectronic device called the mCell. The mCell is a slightly modified MRAM device and acts as a current-controlled switchable resistor. Like MRAM, this device is nonvolatile so the resistance state is memorized even when the controlling current is shut off.

Figure 3 depicts the four-terminal mCell device that consists of a write path (w^+-w^-) and electrically-isolated read path (R-R'). Its write path consists of a low-impedance, ferromagnetic metal connecting the bottom electrodes. These electrodes are magnetic with the moments at the two opposite ends oriented in opposite directions, thereby leaving a domain wall in the write path. When a current is sent through the write path, the electrons become spin-polarized by the source terminal. When the electrons reach a region in the write path of opposing magnetization, a torque is exerted on the local moments that causes them to flip in the direction of the spin polarization, which moves the domain wall in the direction of electron flow. As a result, given the direction of the current, this write path layer can be "programmed" to have magnetization oriented "up" or "down".

The programming of the write path also programs the magnetization of the switchable layer in the read path via exchange coupling through an electrically-insulating magnetic oxide, therby orienting that layer's magnetization parallel or antiparallel to the fixed R and R' terminals. These electrodes sandwich a tunnel barrier, forming an MTJ, and so the coupling results in two stable resistance states. These resistance states are nonvolatile, as removing power has no effect on the orientation of any layer's magnetization. Note that for the device configuration shown (Figure 3), the resistance is actually two tunnel junction resistances in series. However, the tunnel barrier under one of the contacts may be etched through and a purely ohmic contact deposited to lower the resistance (Figure 4).

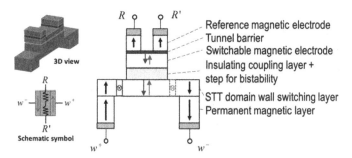

Fig. 3. Schematic symbol, 2D and 3D drawings of mCell.

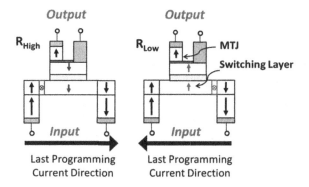

Fig. 4. Programming nonvolatile resistance state is a function on electron flow direction.

Micromagnetic simulations of the mCell solving the Landau-Lifshitz-Gilbert equation modified for STT with current flow in the $\hat{x}$-direction (Eqn. 1) [14] were performed to characterize device behavior. Here, **M** is the magnetization, **H** the effective field, and b_J the STT term, which encompasses the current density.

$$\frac{d\vec{M}}{dt} = -\gamma\vec{M} \times \vec{H} + \alpha\hat{M} \times \frac{d\vec{M}}{dt} - b_J\hat{M} \times \hat{M} \times \frac{\partial\vec{M}}{\partial x} \tag{1}$$

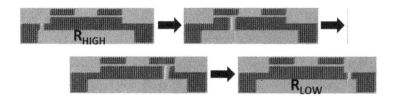

Fig. 5. Micromagnetic simulation of state switching.

A visualization of the state switching process is given in Figure 5, where the mCell switches from a high resistance to a low resistance state with the application of an electron current from the left to right write path terminals.

The cross section in Figure 3 shows that the write path of the mCell has a step to a thicker region. This change in cross-section increases the domain wall energy, and so introduces a significant energy barrier to switching that ensures device bistability and digital switching at room temperature. This is demonstrated in Figure 6, where the domain wall energy leaves its low energy, stable position upon application of a current, and has considerably higher energy as it travels across the write path. In equilibrium, the domain wall naturally rests in the thin necked region on one end of the device.

Additional micromagnetic simulations demonstrate a minimum switching current density of approximately 4 MA/cm^2 with 2 ns pulse width. For a 35nm wide, 3nm thick write path, this current density corresponds to a switching current of roughly 4 µA. In general, higher write currents lead to faster switching times, but the minimum current density required to drive switching tends to level off even at long pulse widths (Figure 7). Simulation demonstrates the domain wall velocity is approximately linear with the current density, as predicted by a first-order model [15]. A Verilog-A compact model of the mCell was written to allow SPICE simulation of circuits, where the base of the model was derived from the linear approximation. Model parameters were then adjusted so that the SPICE simulation would closely match the more accurate micromagnetic model (Figure 8). Simulated current and resistance behavior closely matches measured data from other groups [5], ensuring that the simulation results are reasonable.

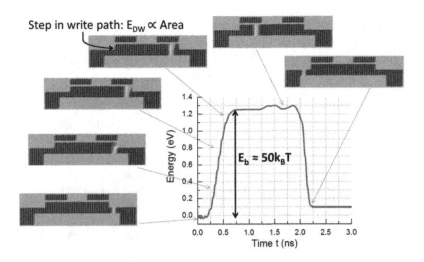

Fig. 6. Energy barrier during switching creates bistable digital device.

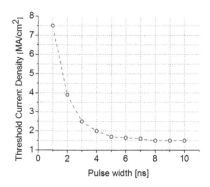

Fig. 7. Micromagnetic simulation results for minimum required current pulse amplitude for switching as a function of pulse width.

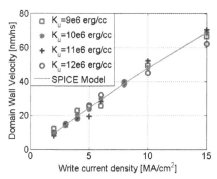

Fig. 8. Matching micromagnetic and SPICE simulated domain wall speed as a function of write current density (parameterized by anisotropy strength).

4. mCell-Based Circuits: mLogic

4.1. *Current Steering*

Switching resistance ratios for MTJs are six to seven orders of magnitude lower than in CMOS devices. With such limited R_{high}/R_{low} ratios, the leakage power of CMOS style circuits increases dramatically. A new approach to logic circuit design is proposed based on current steering. mLogic current-steered circuits can be energy efficient even with switching ratios of 2x-3x. Figure 7 illustrates a simple mLogic buffer with a logic '1' and logic '0' output. The pull-up and pull-down networks are connected to symmetric positive or negative voltages. The output node of this circuit is connected through a low impedance path to ground, represented in Figure 7 as an mCell with write path resistance of 120 ohms (Table 1). The output current is

$$I = \frac{V\left(\frac{1}{R_{PU}} - \frac{1}{R_{PD}}\right)}{2k}, \quad k = 1 + R_{outpath}\left(\frac{1}{R_{PD}} + \frac{1}{R_{PU}}\right) \tag{2}$$

where V/2 is the magnitude of the matching positive and negative power supplies. R_{PU} and R_{PD} are the resistance of the pull-up and pull-down mCells. These resistances are controlled by the input logic value and differ by the R_{high}/R_{low} ratio to steer the output current into or out of the fanout load. The direction of this current is the output logic value.

By design, the magnitude of the output currents representing logic '1' and logic '0' are large enough to ensure writing of the fanout devices in the presence of variation. It is not necessary to have fine control over write current magnitudes as logical value is based on current direction. Margining against variation can simply be achieved by raising power supply voltages which raises the output currents.

In addition to the buffer shown in Figure 9, a variety of other pull-up and pull-down networks can be designed with MTJs to steer the directionality of currents. A NAND gate is shown in Figure 10 MTJ areas are sized so that the gate can steer the necessary currents.

The output current is used to force magnetic state and resistance switching in the fanout mCells. The fanouts are connected in series through their write paths so that each receives the full programming current. The serial connection of fanout also prevents the shunting of current through unbalanced loads in parallel paths.

Table 1. Exemplary Device Parameters

Threshold Current Density [MA/cm^2]	4
MTJ Resistance*Area [ohm*μm^2]	2
Tunnel Magnetoresistance Ratio	100%
Read Path Low Resistance [Ω]	1.25K
Read Path High Resistance [Ω]	2.5K
Write Path Resistance [Ω]	120

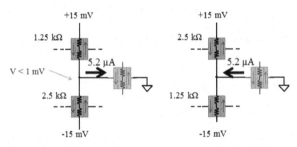

Fig. 9. mLogic steering fanout current direction. Fanout is represented by greyed mCell.

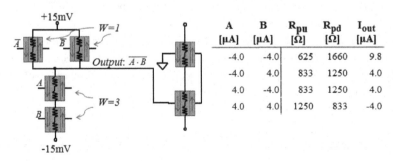

A [μA]	B [μA]	R$_{pu}$ [Ω]	R$_{pd}$ [Ω]	I$_{out}$ [μA]
-4.0	-4.0	625	1660	9.8
-4.0	4.0	833	1250	4.0
4.0	-4.0	833	1250	4.0
4.0	4.0	1250	833	-4.0

Fig. 10. NAND circuit driving fanout inverter and associated truth table. The pulldown mCells are sized three times larger than the pullup mCells.

4.2. Powering and Synchronization

A CMOS transistor in the off-state is nearly an open switch and has little static power consumption. In contrast, current steering with mCell MTJs results in the power supply sourcing 10's of microamps of static current per gate, albeit from sub-100mV supplies. For maximum power efficiency, mLogic gates should not be powered until after each input has received enough current to switch. mLogic gates are non-volatile and do not need to be powered during a write. Similarly, the gates can cease to be powered once their outputs source enough current to switch fanout. Selectively powering the gates is accomplished by strobing the matched positive and negative power supplies. These clocked power supplies are called *pClocks*. When a pClock is enabled, current is steered through the gate to produce an output current to program the fanout.

Figure 11 illustrates the switching of a NAND gate evaluated on the 'A' phase of pClock. A separate non-overlapping pClock powers the fanout inverter. The sign of the output current represents the logical sense of the signal (positive current is a "1" and negative current a "0"). The resistance state switching of a fanout mCell due to this output current is also shown. Note that the resistance state remains stable even between pClock pulses. Like conventional MRAM, the mCells have practically infinite retention time.

The NAND gate and the inverter on its fanout are clocked by separate pClocks. It is preferable for alternating stages of logic to be clocked and powered on non-overlapping phases of pClocks. This is shown in Figure 12 where the three NAND gates in the second stage of logic all are connected to pClk2. In this way, many gates share a common set of pClock signals and as few as two pClocks can be used to power an entire chip.

When none of the pClocks are enabled, the nonvolatile circuit retains state and enters a zero power sleep mode. When the pClock is reenergized, the circuit instantaneously exits the sleep mode and logic, computation continues as normal. No additional overhead is required to support this sleep mode as it would be required in a CMOS chip. pClock signals are distributed globally and shared by many gates, much like VDD and ground are common to all gates in a power domain on a CMOS chip. The global nature of the pClocks allows one to use off-chip power regulation circuitry to generate strobed supplies, thus avoiding the need to tightly integrate CMOS with the magnetoelectronic devices. Two separate die, one for CMOS and one for mLogic, may be used because only limited interconnections are necessary. The interconnections would ideally be made using chip-to-chip bonding, through-silicon-vias, or other 3D integration techniques. Each die is simpler and less expensive to manufacture than one heterogeneous die due to the reduced number of process steps. If a heterogeneous die is preferred, integration is possible as well because the materials and processing are similar to that of conventional MRAM, which is generally considered to be CMOS compatible.

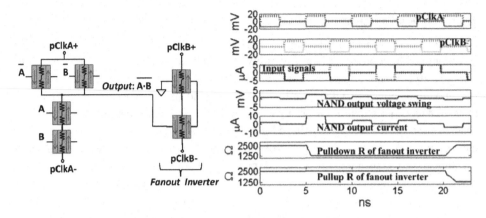

Fig. 11. SPICE simulation of NAND gate driving fanout load (top); evaluation of gate with power clocks, input currents, output current, and resistance state of fanout mCell (bottom).

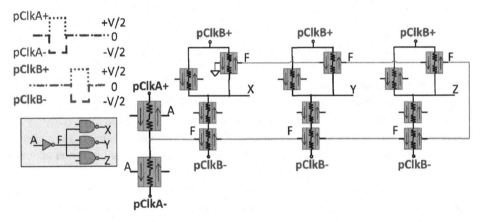

Fig. 12. Inverter steers current through NAND fanout gates; driving gate and fanout powered by non-overlapping pClocks.

5. mLogic Gate Level Behavior

mLogic gates have a number of characteristics that enable efficient implantation of logic functions. Inversion can be done without any additional gates by altering the routing to swap the direction of current on the w^+ and w^- terminals. This is possible because the logical sense of a signal is based on the direction of write current. Free inversion can simplify implementation of logic functions.

For some logic gates, CMOS consumes less energy. For example, a two-input NAND gate with a fanout of four consumes 1.03 fJ/switch in 32 nm CMOS and 1.71 fJ/switch in mLogic (with VDD = 0.9V in CMOS and ±35 mV in mLogic). The 10-20 ps switching speed of a transistor is also significantly higher than that of an mCell. However, these numbers understate the energy advantages of nonvolatile logic. mLogic dissipates significantly less energy in interconnect because with current-based signaling the voltage swing on an output node is generally no greater than a few millivolts. For example, a CMOS inverter consumes 23 fJ to drive 100 µm of metal. This is dominated by a 23 fF load and a tapered buffer chain to drive the load. But an mLogic inverter consumes just 1.4 fJ. The output swings just 3.2 mV with a ±25 mV pClock driving the 700 ohm wire load.

With memory inherent to every mCell, state storage is free. This enables pipelined digital blocks. Simple deep pipelining of mLogic offsets the moderate switching speed of individual mCells to provide high system-level throughput (Figure 13). In CMOS, pipelining registers are generally inserted about every 12 logic stages [16]. In mLogic, it is possible to pipeline every other logic stage, achieving 6x the speedup of coarser pipelining. This aggressive pipelining capability comes with zero additional area, power, and timing overhead. In contrast, a CMOS flip flop may have a timing overhead of three gate delays, use nearly 20 transistors, and consume almost 3 fJ per switch in a 32 nm process.

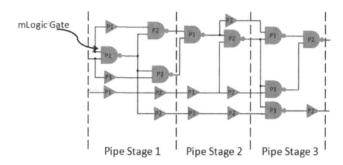

Fig. 13. Logic pipelining.

6. mLogic Architectures

Optimizing logic and architecture for the capabilities of mLogic allow us to obtain system level energy and speed benefits above what is achievable by the gates alone. The ability for each gate to store state is even more significant if this granular memory can be interspersed with the logic, enabling new types of datapaths and architectures. Ultimately, this would necessitate a new approach to logic synthesis. In this section we discuss the specific design of a hardware architecture that takes advantage of the memory state in each mLogic gate to achieve low energy-per-operation at high throughput.

Bit-serial logic styles are particularly suited to exploit the state present in each mLogic gate. Traditionally, adders are implemented with all bits of the summands added concurrently. The critical path of this circuit is generally from the lowest order bit to the highest order bit. In contrast, bit-serial arithmetic [17] operates on a single bit position of the summands per clock and all the bits pass through the same addition gate in sequential order (Figure 14). The carry-out from the addition of one bit position is held in a sequential delay element and fed back into the gate to compute the next bit position.

The advantage of bit-serial logic is that an adder for an arbitrarily wide data word can be fully implemented by one single bit wide adder, yielding significant area savings. Furthermore, multiple adders can be efficiently composed into larger datapaths because only a single wire per adder needs to be routed, instead of a bus as wide as the data word. This arithmetic style is possible in CMOS, but flip-flops and the high timing overhead for clocking relative to the short switching time of a CMOS gate makes this approach inefficient with traditional logic technologies.

Multiplication can also be done compactly using constant coefficient bit-serial logic (Figure 15). Standard constant coefficient multipliers bit-shift the multiplicand to multiply by a power of two and add these products together. mLogic bit-serial bit-shifting is equivalent to a delay element that is as simple as a buffer. Multiplication is equivalent to delay because in the bit-serial representation the bits of the data word pass through the same gates sequentially.

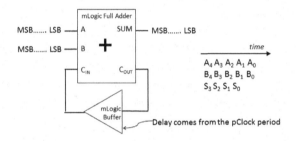

Fig. 14. Bit-serial adder with mLogic.

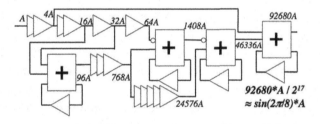

Fig. 15. 16 bit precision bit-serial twiddle multiplier with 22 gates.

Using the building blocks of a bit-serial adder and multiplier, highly efficient FFT hardware can be assembled. The core computation unit of an FFT architecture is the butterfly. This butterfly consists of complex-valued twiddle factor multiplication and an addition and subtraction. In general, FFT hardware architectures use a number of butterfly units to match the application requirements for throughput, power and cost. Higher throughput is enabled by streaming reuse of many duplicated butterfly units per chip while area efficiency is enabled by iterative reuse of fewer butterfly units over several cycles [18].

With mLogic circuits, the FFT hardware tradeoffs between area and latency are recast, and architectures, datapaths, and arithmetic circuits with improved area, throughput and power are possible. Extremely compact pipelined butterfly adders and multipliers can be built with mLogic allowing hardware to efficiently exploit the data parallelism available. The bit-serial datapaths enable an entire 512 point FFT to be implemented without any iterative reuse. This approach would be area inefficient without the sequential bit-serial circuits. With mLogic, the datapath building blocks are simpler and just as importantly the interconnections between the blocks are reduced to single bit wide busses that have swings under 5 mV due to current-based signaling.

All verification of the FFT block was done manually using SPICE on layout extracted netlists. SPICE simulations of the mLogic hardware revealed reduced write currents due to the extracted routing resistances. pClock voltages were raised to compensate. Two voltage domains were used for energy savings, with ±50 mV strobed to the adders and ±20 mV to the buffers.

The simulations showed that bit-serial streaming reuse in mLogic with pipelining allows throughput of 6.5 million FFT/sec at 23.1 nJ/FFT. To put these numbers in context, this is 6.7 times less energy and 1.7 million times the throughput of a seminal sub-threshold CMOS implementation optimized for minimum energy [19]. This increase in throughput comes at just over twice the device count (1.5 million mCells to 627,000 transistors). Sub-threshold CMOS is selected as a point for comparison because like mLogic it is optimized for extremely energy constrained systems.

7. Conclusion

We have proposed a new logic family composed of modified MRAM devices without CMOS integration. These devices switch by spin-transfer torque and have electrically isolated, but magnetically exchange-coupled write path and read MTJ. The isolation enables the MTJs to form resistive networks without integrated CMOS. This new family, mLogic, operates at sub-100 mV. Further, an FFT ASIC shows that new architectures enabled by nonvolatile mLogic can have 6.7 times less energy per operation at 1.7 million times the throughput than a subthreshold CMOS implementation optimized for low energy operation.

References

[1] R. P. Cowburn, "Room Temperature Magnetic Quantum Cellular Automata," *Science*, vol. 287, no. 5457, pp. 1466-1468, Feb. 2000.

[2] A. Imre, G. Csaba, L. Ji, A. Orlov, G. H. Bernstein, and W. Porod, "Majority logic gate for magnetic quantum-dot cellular automata," *Science (New York, N.Y.)*, vol. 311, no. 5758, pp. 205-8, Jan. 2006.

[3] D. A. Allwood, G. Xiong, C. C. Faulkner, D. Atkinson, D. Petit, and R. P. Cowburn, "Magnetic domain-wall logic," *Science (New York, N.Y.)*, vol. 309, no. 5741, pp. 1688-92, Sep. 2005.

[4] D. Allwood et al., "Submicrometer ferromagnetic NOT gate and shift register," *Science (New York, N.Y.)*, vol. 296, no. 5575, pp. 2003-6, Jun. 2002.

[5] S. Fukami et al., "Low-Current Perpendicular Domain Wall Motion Cell for Scalable High-Speed MRAM," in *Symposium on VLSI Technology*, 2009, pp. 2008-2009.

[6] S. S. P. Parkin, M. Hayashi, and L. Thomas, "Magnetic domain-wall racetrack memory," *Science (New York, N.Y.)*, vol. 320, no. 5873, pp. 190-4, Apr. 2008.

[7] M. Hosomi et al., "A Novel Nonvolatile Memory with Spin Torque Transfer Magnetization Switching: Spin-Ram," in *IEEE International Electron Devices Meeting, 2005. IEDM Technical Digest*, 2005, pp. 459-462.

[8] S. Ikeda et al., "Magnetic Tunnel Junctions for Spintronic Memories and Beyond," *IEEE Transactions on Electron Devices*, vol. 54, no. 5, pp. 991-1002, May 2007.

[9] B. Behin-Aein, D. Datta, S. Salahuddin, and S. Datta, "Proposal for an all-spin logic device with built-in memory," *Nature nanotechnology*, vol. 5, no. 4, pp. 266-70, Apr. 2010.

[10] S. S. P. Parkin et al., "Giant tunnelling magnetoresistance at room temperature with MgO (100) tunnel barriers," *Nature materials*, vol. 3, no. 12, pp. 862-7, Dec. 2004.

[11] S. Ikeda et al., "A perpendicular-anisotropy CoFeB-MgO magnetic tunnel junction," *Nature materials*, vol. 9, no. 9, pp. 721-4, Sep. 2010.

[12] S. Matsunaga et al., "MTJ-based nonvolatile logic-in-memory circuit, future prospects and issues," *Design, Automation & Test in Europe Conference & Exhibition, 2009. DATE'09*, pp. 433-435.

[13] F. Ren, S. Member, and D. Markovi, "True Energy-Performance Analysis of the," *Symposium A Quarterly Journal In Modern Foreign Literatures*, vol. 57, no. 5, pp. 1023-1028, 2010.

[14] Z. Li and S. Zhang, "Domain-Wall Dynamics and Spin-Wave Excitations with Spin-Transfer Torques," *Physical Review Letters*, vol. 92, no. 20, pp. 1-4, May 2004.

[15] A. Mougin, M. Cormier, J. P. Adam, P. J. Metaxas, and J. Ferré, "Domain wall mobility, stability and Walker breakdown in magnetic nanowires," *Europhysics Letters (EPL)*, vol. 78, no. 5, p. 57007, Jun. 2007.

[16] W. K. Luk and R. H. Dennard, "2T1D memory cell with voltage gain," *VLSI Circuits, 2004. Digest of Technical Papers. 2004 Symposium on*, 2004.

[17] H. J. Sips, "Bit-Sequential Arithmetic for Parallel Processors," *IEEE Transactions on Computers*, vol. C-33, no. 1, pp. 7-20, Jan. 1984.

[18] P. A. Milder, F. Franchetti, J. C. Hoe, and M. Püschel, "Formal datapath representation and manipulation for implementing DSP transforms," in *Proceedings of the 45th annual conference on Design automation - DAC'08*, 2008, p. 385.

[19] A. Wang and A. Chandrakasan, "A 180-mV subthreshold FFT processor using a minimum energy design methodology," *IEEE Journal of Solid-State Circuits*, vol. 40, no. 1, pp. 310-319, Jan. 2005.

BIOMOLECULAR FIELD EFFECT SENSORS (bioFETs): FROM QUALITATIVE SENSING TO MULTIPLEXING, CALIBRATION AND QUANTITATIVE DETECTION FROM WHOLE BLOOD

ALEKSANDAR VACIC[1] and MARK A. REED[1,2]

Departments of [1]Electrical Engineering and [2]Applied Physics, Yale University
New Haven, CT 06520-8284 USA
[1]*alek.vacic@gmail.com*

Nanoscale field effect based sensors have emerged as a potential label-free diagnostic tool capable of detecting extremely low levels of biomolecules with fast response times. To date, successful detection of various biomolecular species ranging from small molecules to antibodies have been demonstrated, however, the lack of quantitative methods, sensor calibration techniques and ability to detect charges in strong ionic strength environments has hindered their commercial application. In this paper we discuss a recent progress in this field directed primarily towards overcoming the aforementioned obstacles.

Keywords: Label-free; bioFET; ISFET; semiconductor field effect sensor.

Introduction

The ability to sense biomolecules using direct electronic label-free detection in disparate environments is of great interest for biomedical, clinical, pharmaceutical and defense research. Recent advancements in nanofabrication techniques offer a promise for delivering portable electronic platforms capable of rapid, ultrasensitive, low-cost, low-powered and multiplexed identification of various biomolecular species. Compared to the current cutting-edge techniques such as surface plasmon resonance, radio-tags and DNA microarrays, electronic detection allows integration of sensor arrays with data processing components (registers, amplifiers, analog-to-digital converters…), on-chip multiplexing and microfluidic integration. Over the past decade semiconducting nanowires, configured as field effect transistors (FETs), have demonstrated great promise to satisfy these demands and reach sensitivities that are compared to or better than these optical techniques.

The idea of using FETs for detection of biomolecules was first demonstrated three decades ago. An Ion-Sensitive FET (ISFET) (Bergveld and Sibbald, 1988) is similar to a conventional Metal-Oxide-Semiconductor FET (MOSFET), however, instead of a metal gate that is used to turn the device on and off, an ISFET lacks the metal gate, thus allowing the oxide layer to be exposed to the electrolyte solution. This electrolyte-oxide interface contains dangling bonds i.e. binding sites which act as a gate whose electrostatic potential can be modulated by the binding of charged species. Binding of ions (charged molecules, proteins, etc.) from the solution to the ISFET surface causes changes in the semiconductor surface potential, and therefore modulates device current. A typical application of an ISFET for biomolecular detection is achieved by surface modification using enzymes – ENFETs (Schoot and Bergveld, 1988), where bound enzyme catalyzes chemical reaction which causes a local pH change. The major drawback of this technology is a reliance on enzymatic activity which greatly depends on environment conditions (temperature, buffer ionic strength, etc.) and therefore device lifetime and robustness.

Nanowire FETs, first introduced in 2001 (Cui et al., 2001), demonstrated the much desired ability for detecting molecules at clinically relevant levels (Zheng et al., 2005). As demonstrated by semi-empirical approaches (Liu et al., 2003), a generally accepted explanation for the increased sensitivity of nanoscale sensors is due to increased surface-to-volume ratio (Fan and Liu, 2006). Nanoscale FETs based on chemical-vapor deposition (Wagner and Ellis, 1965) CVD- grown nanowires, have been used for detection of small molecules, proteins, biomarkers and viruses, both for complementary and multiplexed sensing. However, significant variation in electrical characteristics (Stern et al., 2005) and difficulties to incorporate CVD grown nanowires in existing top-down fabrication techniques (McAlpine et al., 2007) still represent major shortcomings of this technology.

Improvements in quality of silicon-on-insulator wafers (Bruel, 1995) and processing techniques (Colinge, 1991) gave a significant boost to the use of top-down fabrication techniques. In these processing schemes a device nano-channel is formed by either electron-beam (EBL) or by deep ultraviolet (DUV) lithography, followed by reactive ion etching (RIE) (Pui et al., 2009). Recent experimental studies (Rajan et al., 2010) on 1/f noise of the RIE defined devices have demonstrated a significant degradation of nanowire electrical and transport characteristics such as carrier mobility and subtreshold swing as well as an increased number of traps.

On the other hand, anistropic wet etching of silicon have demonstrated devices with better electrical characteristics (e.g. subthreshold swing, low-frequency noise...) by preserving the sidewall smoothness and lowering the interface trap density (Rajan et al., 2010).

As mentioned before, a very powerful application of nanosensors is direct detection of biomolecules without the need for pre-labeling (i.e. attachment of fluorophores or radiotags). (Patolsky and Lieber, 2005) Detection of specific molecules via direct electronic method requires the sensors to be functionalized with receptor molecules prior to sensing. (Patolsky and Lieber, 2006) The type of target molecules determines the receptor molecules and therefore the surface functionalization scheme.

A major drawback of the bioFET applications is the lack of calibration methods and the inability to quantify biomolecular species. This problem has prevented bioFET technology to become competitive with the current cutting-edge techniques, such as surface plasmon resonance (Jonsson et al., 1991) or ELISA (Enzyme Linked Immunosorbent Assay) (Engvall and Perlmann, 1972). However, several recent works have demonstrated successful multiplexed detection of cancer biomarkers as well as methods for device calibration which would ultimately lead to application of bioFETs for quantitative detection of analytes.

Another handicap for the bioFET technology is the inability or lack of experiments showing label-free detection in environments such as whole blood, plasma or saliva. Frequently reported bio-fouling, non-specific binding, false positive detection, and most critically Debye screening limitations (Stern et al., 2007b, Vacic et al., 2011a) have diminished the competitiveness of the bioFETs. However, recent works have demonstrated two successful approaches. In one, the whole blood is first pre-filtrated using a microfluidic capture-release process which separates target molecules (if present) from the other whole blood components and substitutes them with sensing buffer, therefore circumventing the problems associated with whole blood sensing. The second method utilizes whole blood, but device current is read before injection and after-washing with a sensing buffer.

The main purpose of this article is a review of developments in the bioFET field that have been occurring during the past few years, especially in the arena of multiplexed detection, calibration, quantification and whole blood sensing.

Fabrication

Over the past years several different hybrid and bottom-up methods for nanoscale bioFET fabrication have been demonstrated (Quitoriano and Kamins, 2008, McAlpine et al., 2007). However, two issues are seen as major drawbacks for incorporating these methods in mass production techniques. First and foremost, the use of CVD grown nanowires are incompatible with main stream CMOS technologies which makes integration with on-chip amplifiers and data processing circuits practically impossible. The inability to integrate bottom-up nanowire FET with microelectronic components and microfluidics directly hinders application for multiplexed detection schemes and assays. Finally, the low yield and inability to produce large number of devices with similar electrical characteristics (Stern et al., 2005) prevents analyte quantification which is critical for a competitive point-of-care (POC) diagnostic tool.

Two methods have emerged as a potential solution to this obstacle. Both of them are based on Silicon-On-Insulator as fundamental material, but the main difference is the methods and chemistries used in the fabrication process.

In the pioneering work, Stern et al. (Stern et al., 2007a), introduced a Complementary Metal Oxide Semiconductor (CMOS) compatible nanowire process using Electron Beam Lithography (EBL), Figure 1a. In the first step active silicon layer was thinned to a desired thickness using dry oxidation and buffered-oxide etch. Upon the optical definition of nanowire mesas and back gate via etch, implantation was performed to form source and drain contact followed by activation annealing. Device width was scaled down from micrometer size to several tens of nanometers using EBL. Oxide masks were grown using dry oxidation followed by electron beam photoresist patterning. The crucial step for defining nanowire body is an anisotropic etch using tetramethylammonium hydroxide

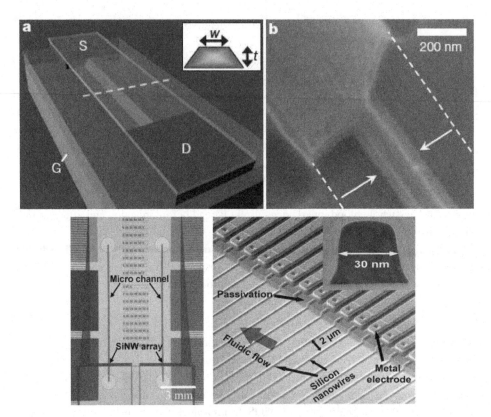

Fig. 1. a) Nanowire bioFET schematic showing characteristic trapezoidal cross-section due to anisotropic etch. b) Scanning electron micrograph of a completed device, (Stern et al., 2007a) c) Optical image of silicon nanowire arrays with integrated microfluidic channel, d) SEM micrograph of silicon nanowire bioFET arrays. The inset shows a TEM cross-section of representative nanowire, (Pui et al., 2009).

(TMAH) which etches (111) planes 100 times slower than the other crystal planes of silicon. This allows precise control of device width as a function of etch time. Due to orientation dependent etch, the cross-section of the nanowires obtain this way is trapezoidal with a characteristic angle of 54.7°, Figure 1b). Finally, devices are then metalized and passivated to protect exposure of metal pads to solution.

The second method (Agarwal et al., 2008) uses deep ultra-violet lithography for patterning nanowire mesas, followed by a dry oxidation to additionally scale down the width to ~80nm, Figure 1c and d. A channel implantation was then performed to adjust the threshold voltage, followed by oxide deposition for nanowire passivation and another implantation for source and drain contacts. Following the metallization and nitride passivation steps, the nanowires are released using an RIE step. This method requires less steps than the original one and, due to the threshold voltage adjustment, the nanowire FETs are always turned on and capable of operating under zero solution gate bias. This is very convenient since it avoids the use of the backgate bias which can additionally increase the diffusion of sodium and potassium ions from solution which cause threshold voltage shift and device degradation over time. On the other hand, the absence of the ability to adjust device operating point and the operation in linear regime limit investigation of device sensitivity as function of bias. According to some reports it was found that nanowire FETs are most sensitive in subthreshold region (Rajan et al., 2011). However, it was also shown that this mode of operation did not have the optimal Signal-to-Noise ratio (SNR), as noise increases in the subthreshold regime, and the optimal SNR operating point is the peak transconductance.

Finally, a very innovative approach was introduced by Elfstrom et al. (Elfstrom et al., 2008), using silicon nanoribbon sensors, which have nanoscale thickness while lateral dimensions are kept on the order of several microns. Preliminary data suggests that the ratio of the Debye screening length in silicon and the electrical thickness of the FET plays critical role for device sensitivity scaling (Reddy et al., 2011).

Whole Blood Detection

Since their introduction in 2001, electronic-label free sensors have demonstrated a range of qualitative applications ranging from proteins (Cui et al., 2001), oligonucleotides (Li et al., 2004), cellular response (Stern et al., 2008), viruses (Patolsky et al., 2004) and cancer biomarkers (Stern et al., 2010b). Despite the huge success in the academic community, these sensors have suffered a fundamental limitation in application due to their inability to operate and sense in physiological solution and, more importantly, whole blood. Debye screening, bio-fouling and non-specific binding are some of the reasons the nanowire bioFET technology has not reached its full potential. Two different approaches can be recognized in the efforts to overcome these obstacles.

A new in-line microfluidic purification chip (MPC) can be integrated with existing bioFETs, Figure 2. The MPC chip (Stern et al., 2010b) operates by capturing analytes of interest from whole blood, exchanging the blood with low ionic concentration sensing buffer and releasing the analytes (specifically in this work, using a photo-cleavable polymer). Using a Bosch etch process (Abdolvand and Ayazi, 2008) the MPC chip is fabricated from bulk silicon wafer and consists of a honeycomb lattice of pillars, 100μm in diameter and 100μm high. These pillars serve to enhance binding surface and therefore increase capture efficiency of the MPC, Figure 2a. Using 3-aminopropyltrietoxy silane (APTES), amine functionality is conferred to the MPC surface which can be further modified with biomolecules using known surface chemistry methods. Following the amine functionalization, the chip was treated with capture antibodies conjugated with the photo-cleavable cross-linker.

The MPC chip operates by introducing whole blood and binding of any potentially present biomarkers to the surface, Figure 2b. The specificity of the MPC is dictated by the type of antibodies on the surface. Upon binding step, the chip is washed using sensing buffer (0.01-0.1×Phosphate Buffer Saline, PBS or 1-10mM bicarbonate buffer). After the washing step, the sensing buffer was left in the chamber and the chip was exposed to ultra-violet (UV) radiation, Figure 2c. Any captured biomarkers are released into the buffer solution together with the capture antibody. The sample is then transferred to a

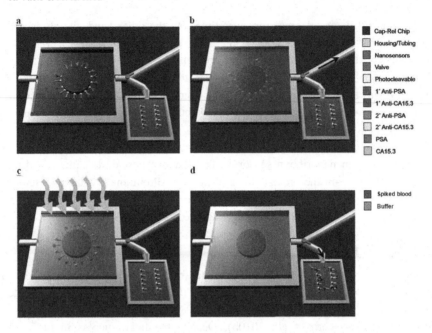

Fig. 2. Microfluidic purification chip (MPC) consisting of honeycomb structure of silicon pillars used by Stern et al. for capture-release filtration of whole blood. a) Primary antibodies are immobilized on the MPC via photocleavablecrosslinker. b) whole-blood is added to the MPC allowing the binding of specific antigens to the MPC surface. Washing step includes substitution of the whole-blood with a buffer solution with low-ionic concentration. c) UV-exposure allows release of immobilized antibody-antigen complexes into the buffer solution. d) in the last step the buffer solution with antibody-antigen complexes is transferred to functionalized nanoribbon chip where the detection occurs, (Stern et al., 2010b).

nanowire/nanoribbon bioFET chip which is functionalized with secondary capture antibodies which bind different epitopes of antigen (biomarker), Figure 2d.

To demonstrate validity of the method as well as the capture effectiveness of the MPC, samples were transferred into a 96-well plate subsequent to cleaving and an enzyme linked immunosorbent assay (ELISA) was performed to determine the concentration of released antibodies. It was calculated, (based on a 10µl sample) that the capture efficiency was ~ 10%. This value is considered good since the whole capture-release process was based on a single drop of blood and very short incubation time (<20 minutes). It is also important to notice that the capture process has not reached saturation, Figure 3, and further improvements in capture efficiency can be achieved by either longer incubation or reflowing the whole blood sample. In addition, the application of UV cleavage step did not affect immunoactivity of antigens and antibodies which was confirmed by ELISA, Figure 3.

Structurally similar MPC approach, but with a different method of operation, was recently introduced. The main principle is a mechanical separation of plasma from whole blood using size-based exclusion (Zhang et al., 2011). By placing submicron vertical pillars in the micro-channels which are in the main flow path of the blood, only plasma is able to pass, Figure 4. To obtain 10ml of plasma sample with 95% efficiency in 15 minutes, it was necessary to achieve 10-15ml/min flow. The more impressive part of this method is that the whole separation chamber was fabricated on the back end of the SOI bioFET chip. Using a deep RIE process the two components were contacted together. This approach could enable cheap portable point-of-care diagnostics, although long-term stability and the effects of blood plasma to bioFET signal stability and reproducibility have not been thoroughly investigated.

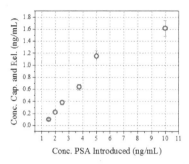 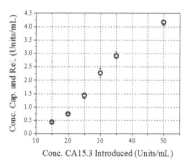

Fig. 3. Scatter plots demonstrating the efficiency of the capture-release process. Concentration of PSA and CA15.3 antigens in a released sample compared to the initial concentrations in the whole blood. Samples were obtained from 10μl of blood in a single 20-minute capture-release process without recirculation of the blood sample, (Stern et al., 2010b).

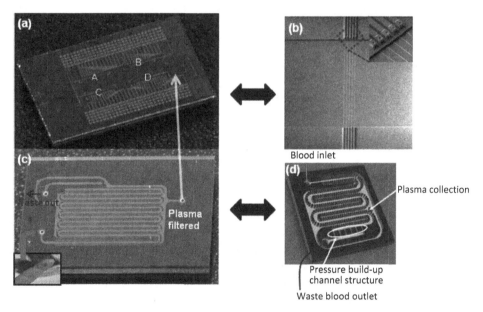

Fig. 4. Nanowire array chip with integrated micropurification system. a) optical image of a nanowire array chip, b) SEM micrographs of nanowires. c) optical image of silicon based filtration chip for whole blood separation. d) Schematic showing working principle behind the filtration chip, (Zhang et al., 2011).

In a method developed by Kim et al. (Kim et al., 2010), Si nanowire arrays were exposed to a reference buffer (10μM phosphate, 20μm NaCl, pH 7.8) and baseline conductance was established. After current stabilization, serum containing cancer biomarkers was passed through the microfluidic channel that allows binding of biomarkers to the sensor surface, Figure 5. Next, reference buffer was flowed to wash the non-bound components, which causes the sensor to reach new current level which is higher than the baseline level (pre-sample injection) due to the presence of bound surface charge. Finally, the baseline signal and post-washing signal were compared and calibration curves were established. Straightforward application of this approach on serum samples without the need for desalting or dilution is a definite advantage; however, issues such as repeatability and reproducibility need to be addressed.

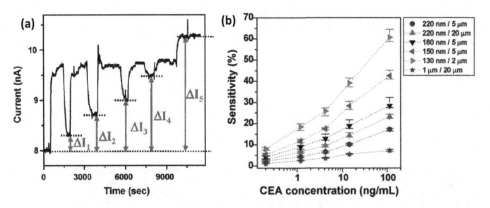

Fig. 5. Direct label-free sensing from human serum. a) Current response versus time of a functionalized silicon bioFET after injection of serum with different concentrations of carcinoembryonic antigen (CEA) followed by a sensing buffer washing step. b) Sensitivity versus CEA concentration obtained from six sensors simultaneously, (Kim et al., 2010).

Calibration and Quantification

So far, significant effort has been made in order to demonstrate qualitative applications of bioFETs in the areas such as detection of biomarkers, oligonucleotides, cellular response, pathogens. In addition, some efforts have been made to suppress device-to-device variations by improving fabrication methods. However, none of these approaches have discussed the possibility of device calibration to suppress non-uniformity of device characteristics or possibilities to use them for quantitative sensing. Moreover, there is no established link between bioFET response and its electrical characteristics.

Most commonly, measurements are done by recording a change in device current ΔI or conductance ΔG before and after addition of analyte. However, measurements have shown both linear and logarithmic dependence on analyte concentration. This has raised discussion about proper operating point and whether sensitivity of a device is larger in the subthreshold rather than the linear region. It can be shown (Chen et al., 2011) that the current change is directly proportional to the change of bioFET surface potential where the proportionality constant is the solution transconductance.

Two papers have already suggested application of solution gating and solution gate transconductance as a mean to calibrate devices (Ishikawa et al., 2009, Stern et al., 2010a). Starting from a MOSFET equation in the linear region:

$$I_{ds} = \mu C_{ox} \frac{w}{L}(V_{GS} - V_T)V_{DS} \tag{1}$$

under an assumption of $V_{DS} \ll V_{GS} - V_T$, the bioFET current change due to the charge binding is:

$$dI_{ds} = -\mu C_{ox} \frac{w}{L}V_{DS}dV_T \tag{2}$$

Normalized sensitivity (i.e. sensor response) is therefore

$$\frac{dI}{I_0} = -\frac{dV_T}{V_{GS} - V_T} \tag{3}$$

while the solution transconductance normalized response is

$$\frac{dI}{g_m} = -dV_T \tag{4}$$

From the last two equations it follows that current normalized response still depends on device threshold voltage and therefore will be affected by any device-to-device variations. The transconductance normalized response is equal to the change in device threshold voltage which is directly related to device surface potential (Bergveld, 1981), therefore by performing transconductance normalization one should expect lower variations in normalized sensitivity for different devices under the same experimental conditions. It is important to notice that previous results are valid only if the total current change (i.e. threshold voltage shifts) happens in the same region of operation. More importantly, if device-to-device variation is low then both current and transconductance normalization yield the same results (Vacic et al., 2011b).

To demonstrate validity of their approach Ishikawa et al. (Ishikawa et al., 2009), applied the transconductance scaling method to a well study system – streptavidin-biotin binding and bottom-up nanowire as a testing platform, which are known to have significant parameter variations. Figure 6 shows responses of three biotin-functionalized devices upon addition of streptavidin. Coefficient of variance (CV), defined as $= \frac{\sigma}{\mu}$, is used as a figure of merit. The CV of raw current changes (ΔI) was around 59% while the baseline normalized CV was reduced to 25%. Further improvement is achieved by transconductance calibration by reducing response spread to 16%.

Equation (4) implies that device responses scaled by corresponding trans-conductances are actually independent on device characteristics and are proportional to

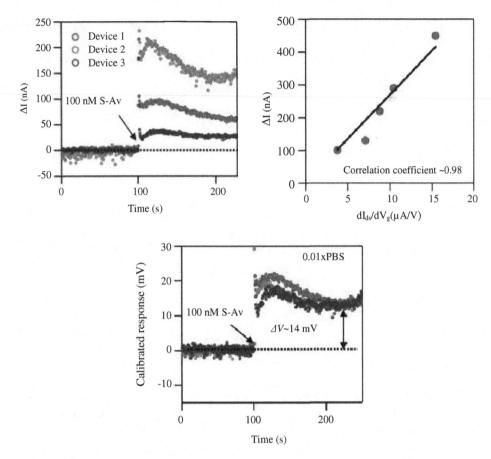

Fig. 6. a) Current versus time response of three biotin-functionalized devices exposed to 100nM streptavidin solution. b) Absolute response dependence as a function of device solution transconductance $g_m = dI_{DS}/dV_{SolG}$. The solid line represents a linear fit with correlation coefficient of 0.98. c) Plots of transconduction normalized responses versus time based on a) and b). Reprinted with permission from Ishikawa et al., 2009. Copyright 2009 American Chemical Society.

threshold voltage shifts. More importantly, according to the basic theory of ISFET operation (Bergveld, 1981) the threshold voltage shift ΔV_T is equivalent to the change in bioFET surface potential $\Delta\psi$. As it can be seen in Figure 6a absolute responses are different for different devices. However, these responses are proportional to device electrical characteristics i.e. transconductance, Figure 6b. Therefore, the transconductance normalized devices would yield a response that is indifferent to device electrical characteristics and only equal to the change in surface potential, assuming that the surface properties and functionalization do no vary significantly form device to device. Using this method one obtains the shift in I_D-V_G characteristics i.e. threshold voltage of approximately 14mV, which is the same for all of the devices used in the experiments discussed below, Figure 6c.

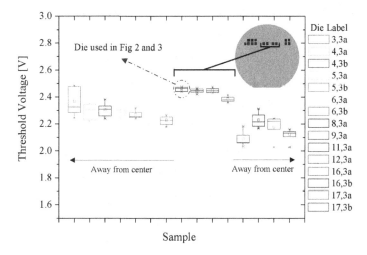

Fig. 7. Average threshold voltage dependence of nanoribbon bioFET as a function of its position on a 4" wafer, (Vacic et al., 2011b).

While these studies have put the field on the right track towards quantitative sensing there are several issues still needed to be addressed. CVD (bottom-up) grown nanowires which were used in these experiments suffer from intrinsic device-to-device variations in mobility threshold voltage and transconductance (Stern et al., 2005). Individual device calibration is therefore required, thus hindering the utilization of multiplexing and integration as major advantages of the microfabrication techniques.

Vacic et al. (Vacic et al., 2011b) further expanded this idea to top-down fabricated devices by demonstrating bioFETs with well controlled threshold voltages of approximately 8mV per die (standard error of the mean, σ_{SEM}) in the areas where the uniformity of the fabrication processes is the highest i.e wafer center, Figure 7. Using this advantage of the top-down processing techniques a novel approach of detection was introduced by measuring initial kinetic (device current) rates rather than the end-point detection which is normally utilized in electronic label-free detection schemes.

The justification for using initial kinetic rates can be found by examining a simple receptor-ligand binding using Langmuir kinetics (Langmuir, 1917). By assuming a reversible reaction $[R] + [L] \leftrightarrow [RL]$ one can calculate time dependence of the relative surface coverage to be

$$R(t) = \left[\frac{k_{on}c}{k_{on}c + k_{off}} - R_0 \right] \left(1 - e^{-(k_{on}c + k_{off})t} \right) + R_0 \qquad (5)$$

and the rate:

$$\frac{dR}{dt} = k_{on}c(1 - R) - k_{off}R \qquad (6)$$

where k_{on} and k_{off} are the association and dissociation constants of the receptor-ligand complex $[RL]$. The change of the relative coverage of the sensor surface at $t = 0^+$ is therefore directly proportional to the analyte concentration provided that the sensor has not been previously utilized i.e. $R(t = 0) = 0$. This fact allows for direct measurement of analyte concentration at the nanosensor surface under assumption that the system is not diffusion limited (Nair and Alam, 2007) and the sensor response is directly proportional to the surface charge (Sorensen et al., 2007, De Vico et al., 2011).

Threshold voltage uniformity of the top-down SOI based bioFETs are primarily affected by the thickness variations of the active silicon layer and the dry oxide used for thinning of the active layer. Figure 7 shows device threshold voltage distribution across 4" wafer demonstrating variations as small as 8mV (<1%). This is an important fact since it allows that all devices on the same chip can be biased at the same operating point, therefore enabling their internal calibration based on the theory expressed by equations (1)-(4).

Figure 8 shows initial current rate dependence of 5 nanoribbon devices as a function of both device baseline and transconductance. As expected according to equations (3) and (4) both dependences are linear. Prior work on bottom-up nanowore FET sensors (reference Ishikawa) have demonstrated better results using transconductance normalization, which is expected according to equation (3) due to the dependence of normalized current change on $V_G - V_T$. On the other hand, transconductance normalization is equal to threshold voltage shift due to the binding of surface charge which is directly related to the change of surface potential $\Delta\psi$ of a bioFET. This is expected for bottom-up nanowires since they exhibit large device-to-device variation in electrical characteristics. For top-down bioFETs this variation can be as low as 8mV (<1%, σ_{SEM}).

Sensing measurements were performed using two different biomarkers – a prostate specific (PSA) and breast cancer (CA15.3) antigens. Measurements were repeated on multiple devices in the clinically relevant concentration range. It was found that the relative standard error of the mean is below 10% on samples obtained from the same stock solution, Figure 9. In addition, a blind measurement tested using a single device on a different measurement setup have demonstrated a good agreement with the calibration curve and concentration obtained using conventional Enzyme Linked Immunosorbent Assay (ELISA).

Further improvement has been seen in work by Zhang et al. (Zhang et al., 2011), who demonstrated simultaneous detection of different biomarkers on a same chip. For certain conditions such as cardiovascular diseases it is necessary to test three critical biomarkers. Standard assays, such as ELISA, are sensitive enough but require skilled and expensive technicians, high consumptions of reagents and are usually performed in centralized institutions which may delay testing. On-chip integration of purification chip and bioFET arrays with appropriate microfluidics may decrease processing time, allow point-of-care diagnostics and lower the cost of such tests. Using standard CMOS compatible processing techniques Zhang et al. integrated the nanowire bioFETs and filtration module

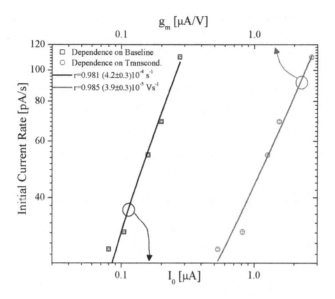

Fig. 8. Initial current rate of five devices from the same die recorded simultaneously as a function of both baseline currents and device solution transconductances at $V_{backgate} = -3V$ and $V_{solution\,gate} = 0V$. The relative standard deviationsfor initial current rates, baseline currents and tranconductances are 0.7%, 0.3% and0.6%, respectively. Both fits are linear (y = kx), shown on a log $-$ log scale for clarity, (Vacic et al., 2011b).

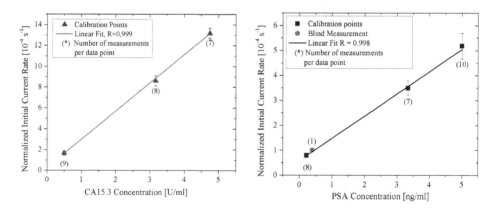

Fig. 9. Calibration curves for (a) PSA and (b) CA15.3 show linear device response in the clinically relevant range of analytes. Red data point represents a blind measurement, (Vacic et al., 2011b).

on the same SOI chip avoiding the need for additional tubing. They demonstrated detection of troponin T (cTnT), creatin kinase MM (CK-MM) and creatin kinase MB (CK-MB) from 2ml of whole blood sample at lowest detection level of 1pg/ml for mere 45 minutes. The filtration chip consisted of an array of pillars with diameters smaller than 0.8mm to allow separation of cell-related components from plasma. After filtration, plasma was flowed over antibody-functionalized nanowires. Upon flowing, excess plasma is washed using a 0.01x Phosphate Buffer Saline (PBS). Nanowire resistance is measured before blood filtration and after washing, both times in PBS. The percentage change in nanowire resistance is directly related to the binding of biomarkers to the nanowire surface. Four groups of nanowire were used in this study three of which were functionalized with appropriate antibodies and one group was functionalized with bovine serum albumin (BSA) as control. Figure 10 shows resistance histogram of multiple devices before and after sensing. As it can be seen there is a significant resistance change for devices functionalized with corresponding antibodies demonstrating the specificity of the antigen-antibody binding. Device functionalized with BSA does not show any significant resistance change for any of the three biomarkers. Even though some non-specific binding and interference from blood plasma was observed it was determined to be 5 to 6 times lower than the real signal.

Detection limit of the proposed methods was tested by lowering the concentrations of biomarkers until sensor response was indistinguishable from the control i.e. blank sample. Measured detection limit is on the order of 1pg/ml which is one order of magnitude lower than the conventional ELISA and 2 magnitudes lower than the commercially available test kits.

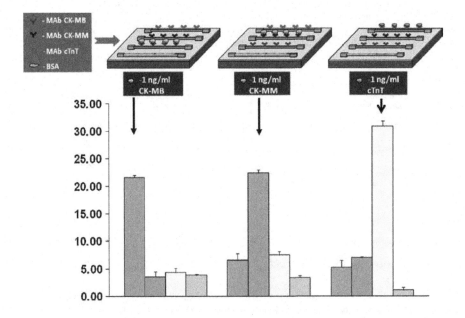

Fig. 10. Specificity of cardiac biomarker detection (cTnT, CK-MM, CK-MB) in plasma using nanowire bioFETs functionalized with appropriate antibodies showing that different biomarkers can be distinguished from the other species in blood. The SiNWs immobilized with BSA exhibit negligible response to the three protein biomarkers, respectively. (Zhang et al., 2011).

Conclusion

To date, numerous papers on silicon field effect nanosensors have been published in scientific journals and conferences. Most of these articles deal with qualitative sensing where FET based sensing platform is applied over and over again to different biological systems varying from protein and oligonucleotide to detection of single viruses and cellular functions. Even though the whole spectrum of applications has successfully been covered, three fundamental topics were never assessed:

1) sensor calibration;
2) quantitative, repeatable and reproducible detection of analytes, and,
3) real-time measurements of biomolecular species in physiological environments, such as whole blood or plasma, which would enable this technology to move forward from academic research labs to diagnostic application.

Recent advancements in fabrication techniques and chip integration have opened-up further improvement of the bioFET technology. So far three approaches have successfully dealt with biomarker detection from whole blood, plasma or serum. The main principle behind these methods is microfluidic separation of whole blood using either a capture-release process or physical (size-based) separation before sensing. More importantly, they mutually complement each other and offer answers to the questions such as calibration, quantification, low detection limit, multiplexing and integration.

All of these challenges have been successfully addressed, and results so far have been very promising. It is very possible that in the near future we would be able to use these devices in our day-to-day life integrated with laptops and even smartphones.

Acknowledgements

Authors would like to acknowledge research support of the National Institute of Health under grant number 1RO1EB008260-1 and the Defense Threat Reduction Agency under grant number HDTRA1-10-1-0037.

References

Abdolvand, R. & Ayazi, F. (2008) An advanced reactive ion etching process for very high aspect-ratio sub-micron wide trenches in silicon. *Sensors and Actuators A: Physical,* 144, 109-116.

Agarwal, A., Buddharaju, K., Lao, I. K., Singh, N., Balasubramanian, N. & Kwong, D. L. (2008) Silicon nanowire sensor array using top-down CMOS technology. *Sensors and Actuators A,* 207-213.

Bergveld, P. (1981) The operation of an ISFET as an Electronic Device. *Sensors and Actuators,* 17-29.

Bergveld, P. & Sibbald, A. (1988) *Analytical and Biomedical Applications of Ion-sensitive Field-Effect Transistors,* Elsevier.

Bruel, M. (1995) Silicon on insulator material technology. *Electronics Letters,* 31, 1201-1202.

Chen, S., Bomer, J. G., Carlen, E. T. & Berg, A. V. D. (2011) Al_2O_3/Silicon NanoISFET with Near Ideal Nersnstian Response. *Nano Lett.,* 6, 2334-2341.

Colinge, J.-P. (1991) *Silicon-on-insulator technology: materials to VLSI.*

Cui, Y., Wei, Q., Park, H. & Lieber, C. M. (2001) Nanowire Nanosensors for Highly Sensitive of Biological and Chemical Species. *Science,* 293, 1289-1292.

De Vico, L., Sorensen, M. H., Iversen, L., Rogers, D. M., Sorensen, B. S., Brandbyge, M., Nygard, J., Martinez, K. L. & Jensen, J. H. (2011) Quantifying signal changes in nano-wire based biosensors. *Nanoscale,* 3, 706-717.

Elfstrom, N., Karlstrom, A. E. & Linnros, J. (2008) Silicon nanoribbons for Electrical Detection of Biomolecules. *Nano Letters,* 8, 945-949.

Engvall, E. & Perlmann, P. (1972) Enzyme-Linked Immunosorbent Assay, Elisa. *The Journal of Immunology,* 109, 129-135.

Fan, Z. & Liu, J. G. (2006) Chemical Sensing With ZnO Nanowire Field-Effect Transistor. *IEEE Transactions on Nanotechnology,* 5, 393-396.

Ishikawa, F. N., Curreli, M., Chang, H.-K., Chen, P.-C., Zhang, R., Cote, R. J., Thompson, M. E. & Zhou, C. (2009) A Calibration Method for Nanowire Biosensors to Suppress Device-to-Device Variation. *ACS Nano,* 3, 3969-3976.

Jonsson, U., Fagerstam, L., Ivarsson, B., Johnsson, B., Karlsson, R., Lundh, K., Lofas, S., Persson, B., Roos, H. & Ronnberg, I. (1991) Real-time biospecific interaction analysis using surface plasmon resonance and a sensor chip technology. *BioTechniques,* 11, 620-627.

Kim, A., Ah, C. S., Park, C. W., Yang, J.-H., Kim, T., Ahn, C.-G., Park, S. H. & Sung, G. Y. (2010) Direct label-free electrical immunodetection in human serum using a flow-through-apparatus approach with integrated field-effect transistors. *Biosensors and Bioelectronics,* 1767-1773.

Langmuir (1917) The constitution and fundamental properties of solids and liquids. II. Liquids. *Journal of the American Chemical Society,* 39, 1848-1906.

Li, Z., Chen, Y., Li, X., Kamins, T. I., Nauka, K. & Williams, R. S. (2004) Sequence-Specific Label-Free DNA Sensors Based on Silicon Nanowires. *Nano Letters,* 4, 245-247.

Liu, Z., Zhang, D., Han, S., Li, C., Tang, T., Jin, W., Liu, X., Lei, B. & Zhou, C. (2003) Laser Ablation Synthesis and Electron Transport Studies of Tin Oxide Nanowires. *Adv. Mater.,* 15, 143-146.

Mcalpine, M. C., Ahmad, H., Wang, D. & Heath, J. R. (2007) Highly ordered nanowire arrays on plastic substrates for ultrasensitive flexible chemical sensors. *Nature Materials,* 6, 379-384.

Nair, P. R. & Alam, M. A. (2007) Design Considerations of Silicon nanowire Biosensors. *IEEE Transactions on Electron Devices,* 54, 3400-3408.

Patolsky, F. & Lieber, C. M. (2005) Nanowire nanosensors. *Materials Today,* 8, 20-28.

Patolsky, F. & Lieber, G. Z. C. M. (2006) Nanowire-based biosensors. *Anal. Chem.,* 78, 4260-4269.

Patolsky, F., Zheng, G., Hayden, O., Lakadamyali, M., Zhuang, Z. & Lieber, C. M. (2004) Electrical detection of single viruses. *Proceedings of the National Academy of Sciences,* 101, 14017-14022.

Pui, T.-S., Agrawal, A., Ye, F., Balasubramanian, N. & Chen, P. (2009) CMOS-Compatible Nanowire Sensor Arrays for Detection of Cellular Bioelectricity. *Small,* 5, 208-212.

Quitoriano, N. J. & Kamins, T. I. (2008) Integratable Nanowire Transistor. *Nano Letters,* 8, 4410-4414.

Rajan, N. K., Routenberg, D. A., Chen, J. & Reed, M. A. (2010) 1/f Noise of Silicon Nanowire BioFETs. *Elect. Dev. Lett.,* 31, 615-617.

Rajan, N. K., Routenberg, D. A. & Reed, M. A. (2011) Optimal signal-to-noise ratio for silicon nanowire biochemical sensors. *Applied Physics Letters,* 98, 3.

Reddy, B., Dorvel, B., Go, J., Nair, P., Elibol, O., Credo, G., Daniels, J., Chow, E., Su, X., Varma, M., Alam, M. & Bashir, R. (2011) High-k dielectric Al_2O_3 nanowire and nanoplate field effect sensors for improved pH sensing. *Biomedical Microdevices,* 13, 335-344.

Schoot, B. H. V. D. & Bergveld, P. (1988) ISFET Based Enzyme Sensors. *Biosensors*, 161-186.

Sorensen, M. H., Mortensen, N. A. & Brandbyge, M. (2007) Screening model for nanowire surface-charge sensors in lquid. *Applied Physics Letters.*

Stern, E., Cheng, G., Chimpoiasu, E., Klie, R., Guthrie, S., Klemic, J. F., Kretzschmar, I., Steinlauf, E., Turner-Evans, D. B., Broomfield, E., Hyland, J., Koudelka, R., Boone, T., Young, M., Sanders, A., Munden, R., Lee, T., Routenberg, D. A. & Reed, M. A. (2005) Electrical Characterization of single GaN nanowires. *Nanotechnology*, 16, 2941-2953.

Stern, E., Klemic, J. F., Routenberg, D. A., Wyrembak, P. N., Turner-Evans, D. B., Hamilton, A. D., Lavan, D. A., Fahmy, T. M. & Reed, M. A. (2007a) Label-free immunodetection with CMOS-compatible semiconducting nanowires. *Nature*, 519-522.

Stern, E., Steenblock, E. R., Reed, M. A. & Fahmy, T. M. (2008) Label-free Electronic Detection of the Antigen-Specific T-Cell Immune Response. *Nano Letters*, 8, 3310-3314.

Stern, E., Vacic, A., Li, C., Ishikawa, F. N., Zhou, C., Reed, M. A. & Fahmy, T. M. (2010a) A Nanoelectronic Enzyme-Linked Immunosorbent Assay for Detection of Proteins in Physiological Solutions. *Small*, 6, 232-238.

Stern, E., Vacic, A., Rajan, N. K., Criscione, J. M., Park, J., Ilic, B. R., Mooney, D. J., Reed, M. A. & Fahmy, T. M. (2010b) Label-free biomarker detection from whole blood. *Nature Nanotechnology*, 138-142.

Stern, E., Wagner, R., Sigwort, F. J., Breaker, R., Fahmy, T. M. & Reed, M. A. (2007b) Importance of the Debye Screening Length on Nanowire Field Effect Tranistor Sensors. *Nano Letters*, 7, 3405-3409.

Vacic, A., Criscione, J. M., Rajan, N. K., Stern, E., Fahmy, T. M. & Reed, M. A. (2011a) Determination of Molecular Configuration by Debye Length Modulation. *Journal of the American Chemical Society*, 133, 13886-13889.

Vacic, A., Criscione, J. M., Stern, E., Rajan, N. K., Fahmy, T. & Reed, M. A. (2011b) Multiplexed SOI BioFETs. *Biosensors and Bioelectronics*, 28, 239-242.

Wagner, R. S. & Ellis, W. C. (1965) the Vapor-liquid-Solid Mechanism of Crystal Growth and Its Application to Silicon. *Transactions of the Metallurgical Society of AIME*, 233.

Zhang, G.-J., Luo, Z. H. H., Huang, M. J., Ang, J. A. J., Kang, T. G. & Ji, H. (2011) An integrated chip for rapid, sensitive, and multiplexed detection of cardiac biomarkers from fingerprick blood. *Biosensors and Bioelectronics*, 28, 459-463.

Zheng, G., Patolsky, F., Cui, Y., Wang, W. U. & Lieber, C. M. (2005) Multiplexed electrical detection of cancer markers with nanowire sensor arrays. *Nature Biotechnology*, 23, 1294-1301.

THEORETICAL INVESTIGATION OF INTRABAND, INFRARED ABSORBANCE IN INORGANIC/ORGANIC NANOCOMPOSITE THIN FILMS WITH VARYING COLLOIDAL QUANTUM DOT SURFACE LIGAND MATERIALS

KEVIN R. LANTZ

Electrical and Computer Engineering Department, Duke University, Hudson Hall Room 129
Durham, NC 27713, United States
kevin.lantz@duke.edu

ADRIENNE D. STIFF-ROBERTS

Electrical and Computer Engineering Department, Duke University, Hudson Hall Room 129
Durham, NC 27713, United States
adrienne.stiffroberts@duke.edu

Hybrid nanocomposite thin films, composed of inorganic colloidal quantum dots (CQDs) embedded in a matrix of organic conjugated polymer, have shown promise as a method for room-temperature infrared detection due to the three-dimensional confinement of the CQD and significantly lower dark currents compared to inorganic detectors. However, in order to improve device performance, the excited charges must be efficiently promoted out of the CQD, which is surrounded by an insulating surface ligand. These short, organic molecules, which are required to prevent agglomeration of CQDs in solution, have been shown to inhibit charge transfer into and out of the CQD. In this work, the transfer matrix method is utilized to calculate the quantized energy levels and wavefunctions in the conduction band of the CdSe CQD for a variety of surface ligand materials. These results are used to calculate the absorption coefficient for a size distribution of CQDs and are compared with measured Fourier Transform Infrared absorbance spectra. Finally, the effect of the ligand on the calculated absorption coefficient will be used to optimize the design for an infrared photoconductor.

Keywords: Infrared detection; colloidal quantum dot; conjugated polymer; transfer matrix method.

1. Introduction

The ability to detect infrared (IR) radiation is vital for a host of applications in the fields of medicine, military, atmospheric science, and astronomy. Colloidal quantum dots (CQDs) show promise as a method for achieving near room-temperature infrared detection due to quantum confinement effects. Specifically, as the size of the semiconductor material is reduced below the Bohr radius the conduction and valence bands resolve into single excitonic states and the electron-hole pairs become tightly bound excitons. This three-dimensional confinement of charge carriers in the CQD significantly decreases the dark current in photodetector devices and allows for high temperature operation. Colloidal quantum dots have a distinct advantage over quantum dot materials grown epitaxially in that they are chemically synthesized[1,2] and made soluble in a host of organic solvents through the inclusion of a surface ligand material.

Surface ligands are comprised of short, organic molecules that coat the surface of the CQD and prevent agglomeration of the CQDs when placed in solution. The CQD solution can then be mixed with a conjugated polymer in solution to form a nanocomposite that can be deposited on a substrate through a variety of methods including drop-casting, spin-casting, ink jet printing, pulsed laser deposition, and matrix-assisted pulsed laser evaporation (MAPLE). The conjugated polymer serves two purposes in the nanocomposite films: i) providing a potential offset from the CQD, which gives rise to quantized states; ii) providing a pathway for transport of free carriers excited out of the CQDs. Figure 1 shows a schematic of the energy band diagram for a CdSe/octadecylamine (ODA) surface ligand/poly[2-methoxy-5-(2'-ethylhexyloxy)-1,4-(1-cyanovinylene)phenylene] (MEH-CN-PPV) nanocomposite. While the exact value for the ligand potential offset is unknown, the electron affinity of the ligand material is much lower than the CQD and conjugated polymer and is assumed to approximate the vacuum level.

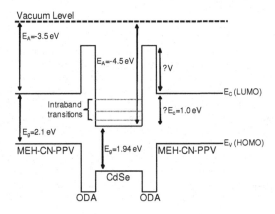

Fig. 1. Schematic diagram of the energy band lineup for CdSe/MEH-CN-PPV nanocomposite thin film where the CdSe CQD is surrounded by the surface ligand octadecylamine (ODA).

Inorganic/organic nanocomposite materials have been increasingly used for many optoelectronic devices including light emitting diodes[3,4], photovoltaics[5,6], and IR photodetectors. To date, the majority of IR photodetectors that have been reported using CQDs[1,7-8] or nanocomposite thin films have demonstrated detection in the near-IR regime (1-3 μm) corresponding to the CQD bandgap energy[9-12]. These devices rely on bipolar, interband transitions (Fig. 2(a)) and photoconduction occurs by hole transport through the conjugated polymer and/or electron transport by hopping conduction through the CQDs. The unique focus of this research is to push the detection into the mid- (3-5 μm) to –long wave (8-12 μm) IR regimes through the use of intraband transitions in the conduction band of the CQD as seen in Fig. 2(b). In these unipolar devices, the conjugated polymer is chosen so that a type-II band lineup is achieved in the CQD, blocking holes from entering confined states, and electron conduction occurs through the conjugated polymer as well as by hopping conduction in the CQDs. It is important to note that the effect of the surface ligand material on charge transport has not typically been considered in devices due to their small size compared to the CQD.

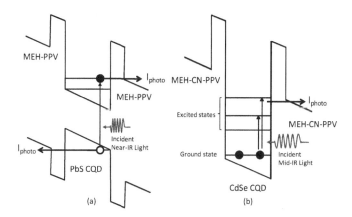

Fig. 2. Schematic representation of IR photodetection via (a) interband transitions across the bandgap of PbS CQDs embedded in MEH-PPV and (b) intraband transitions in the conduction band of CdSe CQDs embedded in MEH-CN-PPV.

Infrared, intraband absorption in CdSe/ODA/MEH-CN-PPV nanocomposite thin films drop-cast on GaAs substrates has previously been measured[13]. Furthermore, an IR photoconductor device using this material system has demonstrated a room-temperature spectral response peak at 5.4 μm (0.228 eV)[14]. However, in order to improve device performance, the excited electrons must be efficiently promoted out of the CQD and through the surface ligand, which acts like an insulative barrier. The choice of surface ligand has been shown to have a profound effect on the transfer efficiency of charges into and out of the CQD[15], and is, therefore, an important optimization parameter that must be understood.

In this work, the effect of the surface ligand material on the quantized energy levels, wavefunctions, and intraband absorption of a CdSe/MEH-CN-PPV nanocomposite is investigated. The results of this model are compared with measured Fourier Transform IR absorbance spectra and used to determine the effect of a variety of surface ligand materials on the calculated absorption coefficient.

2. Theory

2.1. *Calculating the Energy Levels and Radial Wavefunction*

The radial potential profile for the nanocomposite is shown in Fig. 3.

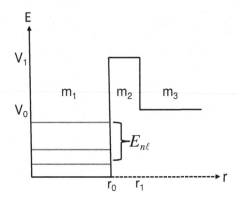

Fig. 3. Schematic representation of finite spherical potential for CdSe/surface ligand/MEH-CN-PPV nanocomposite.

The radial and orbital wave functions that satisfy the Schrödinger equation for this potential are given as

$$R_{n\ell}(r) = \begin{cases} A_\ell j_\ell(kr) + B_\ell n_\ell(kr) & r \leq r_0 \\ C_\ell h_\ell^{(1)}(i\kappa r) + D_\ell h_\ell^{(2)}(i\kappa r) & r_0 < r < r_1 \\ F_\ell h_\ell^{(1)}(i\kappa r) + G_\ell h_\ell^{(2)}(i\kappa r) & r > r_1 \end{cases}, \tag{1}$$

$$Y_\ell^m(\theta,\varphi) = \varepsilon\sqrt{\frac{(2\ell+1)}{4\pi}\frac{(\ell-|m|)!}{(\ell+|m|)!}}e^{im\varphi}P_\ell^m(\cos\theta)$$

where n, ℓ, and m are quantum numbers, j_ℓ and n_ℓ are spherical Bessel and Neumann functions, $h_\ell^{(1)}$ and $h_\ell^{(2)}$ are spherical Hankel functions of the first and second kind, $\varepsilon = (-1)^m$ for $m \geq 0$ and $\varepsilon = 1$ for $m < 0$, $P_\ell^m(\cos\theta)$ is the associated Legendre function, and k and κ are defined as:

$$k = \sqrt{\frac{2m^*E}{\hbar^2}} \tag{2}$$

$$\kappa = \sqrt{\frac{2m^*(E-V(r))}{\hbar^2}}.$$

By applying the boundary conditions that the radial wavefunction and its derivative be continuous for the qth boundary a relationship can be derived between the coefficients in the first region with those in the last region:

$$\begin{pmatrix} F_\ell \\ G_\ell \end{pmatrix} = \begin{pmatrix} M_{11} & M_{12} \\ M_{21} & M_{22} \end{pmatrix}\begin{pmatrix} C_\ell \\ D_\ell \end{pmatrix} = \begin{pmatrix} M_{11} & M_{12} \\ M_{21} & M_{22} \end{pmatrix}\begin{pmatrix} m_{11} & m_{12} \\ m_{21} & m_{22} \end{pmatrix}\begin{pmatrix} A_\ell \\ B_\ell \end{pmatrix}. \tag{3}$$

These relationships can be refined further by forcing the radial wavefunction to go to zero as r goes to infinity and to be well behaved near the origin, such that $G_\ell = B_\ell = 0$. With this constraint, Eq. (3) has only one non-trivial solution:

$$M_{21}(E) = 0 \,, \tag{4}$$

which yields the quantized energy states in the CQD. These energy states are then utilized to solve for the radial wavefunctions in all three regions.

For this work, the energy levels and radial wavefunctions are calculated for the first three orbital angular momentum numbers ($\ell = 0,1,2$) as the contribution to the absorption coefficient is negligible for higher values. Furthermore, only quantized energy levels below the polymer potential offset (V_0) are considered.

2.2. *Calculating the Absorption Coefficient*

The transition rate from an initial to a final energy state due to the absorption of a photon is given by the Fermi golden rule:

$$W_{i \to f} = \frac{2\pi}{\hbar} \left| M_{if} \right|^2 \delta\left(E_f - E_i - \hbar\omega \right), \tag{5}$$

where $\hbar\omega$ is the energy of the absorbed photon and M_{if} is the interaction matrix element. The delta function in Eq. (5) can be replaced by a Lorentzian function, $L(E)$, to model the finite lifetime of an electron in the excited state.

The interaction matrix in Eq. (5) for an incident plane wave is given as:

$$M_{if} = \left\langle \Psi_f \left| \frac{e}{m^* c} \vec{A} \cdot \vec{p} \right| \Psi_i \right\rangle = \frac{e}{m^* c} A_0 \left\langle \Psi_f \left| e^{i\vec{k} \cdot \vec{r}} \hat{\varepsilon} \cdot \vec{p} \right| \Psi_i \right\rangle, \tag{6}$$

where A_0 is the electric field strength, $\hat{\varepsilon}$ is the polarization direction, and $\Psi_{n\ell m} = R_{n\ell} Y_\ell^m$. This equation can be further simplified by employing the electric dipole approximation, such that:

$$M_{if} = \frac{e}{m^* c} A_0 \left(\frac{i m^* \left(E_f - E_i \right)}{\hbar} \right) \langle R_{n\ell,f} \left| \hat{\varepsilon} \cdot \vec{r} \right| R_{n\ell,i} \rangle \langle Y_{\ell,f}^m \left| \hat{\varepsilon} \cdot \vec{r} \right| Y_{\ell,i}^m \rangle. \tag{7}$$

The integration of the radial wavefunctions in Eq. (7) is calculated using the computer model and the integration of the orbital wavefunctions gives rise to the transition selection rules, such that $\Delta\ell = \pm 1$ and $\Delta m = 0, \pm 1$.

For this work, the contribution to the absorption coefficient from all possible energy transitions for a CQD with radius r_0 as well as the contribution from a Gaussian size distribution, $g(r_0)$, of CQDs is compared, such that the absorption coefficient can be defined as:

$$\alpha(\hbar\omega) = \sum_{r_0} \sum_{i,f} \frac{1}{\hbar\omega} \left(\frac{4\pi^2 \left(E_f - E_i \right)^2}{\left(4/3 \, \pi r_0^3 \right) \hbar c} \right) \left| \langle \Psi_f \left| \hat{\varepsilon} \cdot \vec{r} \right| \Psi_i \rangle \right|^2 L(E) g\left(r_0 \right). \tag{8}$$

3. Results and Discussion

3.1. *Comparison of Model with Measured Absorbance*

In order to determine the efficacy of this model the calculated absorption coefficient is compared with the measured absorbance spectrum for a CdSe/ODA/MEH-CN-PPV nanocomposite thin film measured by Fourier Transform Infrared spectroscopy. The absorbance spectrum for the nanocomposite thin film drop-cast on a p-type GaAs substrate is shown in Fig. 4. The details of the measurement and identification of peaks are described elsewhere[13]; however, the peaks at ~0.22 and ~0.43 eV were determined to be due to intraband absorption in the conduction band of the CdSe CQDs.

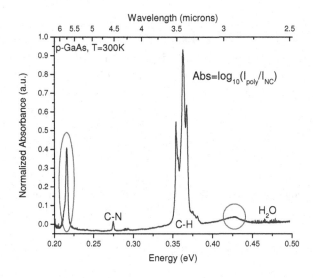

Fig. 4. Absorbance spectrum for CdSe/ODA/MEH-CN-PPV nanocomposite thin film drop-cast on p-GaAs. The circled peaks are absorption due to intraband transitions in the CdSe CQD.

When comparing the calculated absorption coefficient with the measured CQD absorbance peaks the parameters shown in Table 1 are used. The ligand length for ODA is determined by the C-C (1.54 Å) and C-N (1.47 Å) bond lengths, the effective mass for the CQD is that of bulk CdSe, the potential offset for MEH-CN-PPV is taken from literature[16], the assumed values for the mean CQD radius and size distribution are provided by the manufacturer, NN Labs Inc., and the effective mass for MEH-CN-PPV is assumed to be that of the free electron[17]. The mean CQD radius was allowed to vary over a small range around the published value and the unknown variables were used to fit the calculated absorption coefficient to the measured intraband peaks identified in Fig. 4. The comparison of the calculated and measured data is shown in Fig. 5 along with the fitting parameters.

Table 1. Parameters used to fit calculated absorption coefficient with measured absorbance.

Known Variables	Assumptions	Unknowns
r_1-r_0=2.26 nm m_1=m_{CdSe}=0.119m_0 V_0=1.0 eV	<r_0>~3.4 nm Δr_0=0.05r_0 m_3=m_0	m_2 V_1

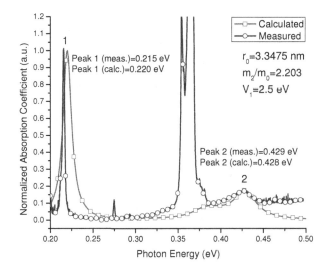

Fig. 5. Comparison of normalized calculated absorption coefficient and measured absorbance with Γ_E=25 meV.

From this figure it can be seen that the relative peak heights fit very well and the peak energies correspond well with a percent error for peak 1 of 2.27% and for peak 2 of 0.11%. It should be noted that the full-width half-maximum (FWHM) of the calculated peak 1 is much larger than that of the measured peak 1 due to the choice of a constant energy spread in $L(E)$ for all excited states (Γ_E = 25 meV). The agreement in the FWHM for peak 1 can be improved by allowing the energy spread for excited states to vary, which has a greater effect on the FWHM of peak 1 since it is the dominant peak for CQD radii near the mean radius, whereas peak 2 tends to dominate for radii further from the mean radius and their contribution is limited by the Gaussian function. Furthermore, the energy of peak 1 matches very closely to the peak responsivity energy (0.228 eV) demonstrated by an IR photoconductor device described in a previous work[14].

3.2. *Comparison of Ligand Materials*

The effect of various ligands on the calculated absorption coefficient, assuming the same material parameters for the CdSe CQD and MEH-CN-PPV polymer, is of interest as a method for optimizing device performance. When comparing ligand materials, the pertinent parameters are the ligand length (r_1-r_0), ligand potential offset (V_1), and ligand effective mass (m_2). The ligand length is determined by the bond lengths for C-C (1.54Å), C-N (1.47Å), C=C (1.34Å), and C-O (1.43Å); however, the lack of values for the potential offset and effective mass in the literature leads to the assumption that these values are the same as those from the fit shown in Fig. 5. This assumption is reasonable because the ligand structures are similar and, thus, the material parameters should not change appreciably from ligand to ligand. The ligand materials and parameters compared in this work are shown in Table 2 and the normalized absorption coefficient for each ligand material is shown in Fig. 6.

Table 2. Model parameters for surface ligand materials.

Surface Ligand Material	Ligand Length (nm)	V_1 (eV)	m_2/m_0
Pyridine	0.308	2.5	2.203
Butylamine	0.497	2.5	2.203
Octylamine	1.000	2.5	2.203
Dodecylamine	1.378	2.5	2.203
Oleic Acid	2.000	2.5	2.203
Octadecylamine	2.258	2.5	2.203

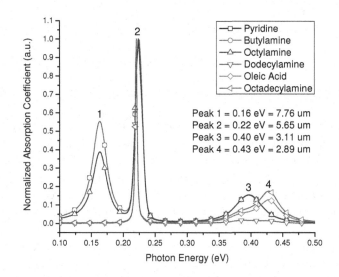

Fig. 6. Comparison of normalized absorption coefficient for various ligand materials with $\Gamma_E = 25$ meV.

From this figure it can be seen that, for the shortest ligand materials, peak 4 is suppressed and peaks 1 and 3 are enhanced. As the ligand length increases, peak 1 disappears and peak 4 becomes stronger, whereas for all ligand materials the strongest peak occurs at ~0.23 eV. The normalized absorption coefficient can be utilized to determine where a device would be sensitive in the IR, however, it is also important to consider how the absorption strength changes with an increase in the ligand length. Figure 7 shows how the absorption coefficient values change for each surface ligand material. It should be noted that, for comparison, the absorption coefficient for epitaxially grown InAs/GaAs quantum dots is ~10^{-4} cm^{-1} [18].

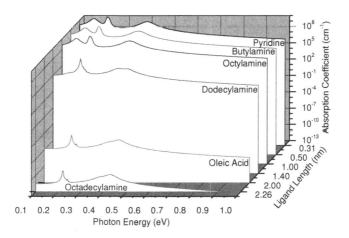

Fig. 7. Comparison of absorption coefficient for various ligand materials with $\Gamma_E = 25$ meV.

From this figure it can be seen that the absorption strength of the nanocomposite material is extremely sensitive to the length of the surface ligand material. This can have a profound effect on the responsivity of a device and it is clear that utilizing the shortest ligand material possible is optimal for improving device performance.

4. Conclusions

In this work a computer model based on the transfer matrix method was developed to model the electronic structure, wavefunctions, and absorption coefficient for a size distribution of CdSe CQDs coated with various surface ligand materials and embedded in the conjugated polymer MEH-CN-PPV. The calculated results of this model were fit to FTIR absorbance measurements of a CdSe/ODA/MEH-CN-PPV nanocomposite thin film and the fitting parameters were used to compare the effect of surface ligand length on the absorption coefficient. This model is the first to realistically calculate the effect of the surface ligand on the optical properties of a nanocomposite system and is an effective design and optimization tool for the fabrication of nanocomposite IR photodetectors. Further work is warranted to determine the bandgap and effective mass of technologically interesting ligand materials to create the most realistic model possible of the nanocomposite system.

5. Acknowledgements

This work was supported by the Office of Naval Research under Grant No. N00014-10-1-0481.

References

1. C. B. Murray, S. Sun, W. Gaschler, H. Doyle, T. A. Betley, and C. R. Kagan, Colloidal synthesis of nanocrystals and nanocrystal superlattices, *IBM J. Res. & Dev.* **45**, 47–56 (2001).
2. C. B. Murray, D. J. Norris, and M. G. Bawendi, Synthesis and characterization of nearly monodisperse CdE (E = sulfur, selenium, tellurium) semiconductor nanocrystallites, *J. Am. Chem. Soc.* **115**, 8706–8715 (1993).
3. V. L. Colvin, M. C. Schlamp, and A. P. Allivisatos, Light emitting diodes made from cadmium selenide nanocrystals and a semiconducting polymer, *Nature*, **370**, 354–357 (1994).
4. H. Mattousi, L. H. Radzilowski, B. O. Daccousi, E. L. Thomas, M. G. Bawendi, and M. F. Rubner, Electroluminescence from heterostructures of poly(phenylen vinylene) and inorganic CdSe nanocrystals, *J. App. Phys.*, **83**, 7965–7974 (1998).
5. D. Cui, J. Xu, T. Zhu, G. Paradee, S. Ashok, and M. Gerhold, Harvest of near infrared light in PbSe nanocrystal-polymer hybrid photovoltaic cells, *App. Phys. Lett.*, **88**, 183111–1–3 (2006).
6. S. A. McDonald, G. Konstantatos, S. Zhang, P. W. Cyr, E. J. D. Klem, L. Levina, and E. H. Sargent, Solution-processed PbS quantum dot infrared photodetectors and photovoltaics, *Nature*, **4**, 138–143 (2005).
7. A. P. Alivisatos, Perspectives on the Physical Chemistry of Semiconductor Nanocrystals, *J. Phys. Chem.*, **100**, 13226–13239 (1996).
8. W. W. Yu, L. Qu, W. Guo, and X. Peng, Experimental Determination of the Extinction Coefficient of CdTe, CdSe, and CdS Nanocrystals, *Chem. Mater.* **15**, 2854–2860 (2003).
9. K. R. Choudhury, W. J. Kim, Y. Sahoo, K.-S. Lee, and P. N. Prasad, Solution-processed pentacene quantum-dot polymeric nanocomposite for infrared detection, *Appl. Phys. Lett.* **89**, 051109–1–3 (2006).
10. G. Konstantatos, I. Howard, A. Fischer, S. Hoogland, J. Clifford, E. Klem, L. Levina, and E. H. Sargent, Ultrasensitive solution-cast quantum dot photodetectors, *Nature* **442**, 180–182 (2006).
11. X. Ma, B. Yuan, and Z. Yan, The photodetector of Ge nanocrystals/Si for 1.55 μm operation deposited by pulsed laser deposition, *Opt. Commun.* **260**, 337–339 (2006).
12. G. Konstantatos and E. H. Sargent, Solution-Processed Quantum Dot Photodetectors, *Proc. IEEE* **97**, 1666–1683 (2009).
13. A. D. Stiff-Roberts and K. R. Lantz, Room-temperature, intraband, infrared absorption in CdSe/poly[2-methoxy-5-(2′-ethylhexyloxy)-1,4-(1-cyanovinylene)phenylene] nanocomposites drop cast on GaAs, *J. App. Phys.* **103**, 104316–1–8 (2008).
14. A. D. Stiff-Roberts, K. R. Lantz, and R. Pate, Room-temperature, mid-infrared photodetection in colloidal quantum dot/conjugated polymer hybrid nanocomposites: a new approach to quantum dot infrared photodetectors, *J. Phys. D: Appl. Phys.* **42**, 234004–1–11 (2009).
15. T-W. F. Chang, S. Musikhin, L. Bakueva, L. Levina, M. A. Hines, P. W. Cyr, and E. H. Sargent, Efficient excitation transfer from polymer to nanocrystals, *App. Phys. Lett.* **84**, 4295–4297 (2004).
16. C. E. Finlayson, D. S. Ginger, E. Marx, and N. C. Greenham, Electrical and optical properties of semiconductor nanocrystals, *Trans. R. Soc. London, Ser. A* **361**, 363–377 (2002).
17. S. J. Martin, J. M. Lupton, I. D. W. Samuel, and A. B. Walker, Modelling temperature-dependent current-voltage characteristics of an MEH-PPV organic light emitting device, *J. Phys.: Condens. Matter* **14**, 9925–9933 (2002).
18. J.-Z. Zhang and I. Galbraith, Intraband absorption for InAs/GaAs quantum dot infrared photodetectors, *Appl. Phys. Lett.* **84**, 1934–1936 (2004).

AUTHOR INDEX